THE
COMPASSIONATE
BRAIN

THE COMPASSIONATE BRAIN

—

HOW EMPATHY CREATES INTELLIGENCE

—

Gerald Hüther

Translated by Michael H. Kohn

TRUMPETER

Boston & London

2006

Trumpeter Books
An imprint of Shambhala Publications, Inc.
Horticultural Hall
300 Massachusetts Avenue
Boston, Massachusetts 02115
www.shambhala.com

9 8 7 6 5 4 3 2 1

First Edition
Printed in the United States of America

Designed by Ruth Kolbert

∞ This edition is printed on acid-free paper that meets
the American National Standards Institute z39.48 Standard.
Distributed in the United States by Random House, Inc.,
and in Canada by Random House of Canada Ltd

Library of Congress Cataloging-in-Publication Data

Hüther, Gerald.
[Bedienungsanleitung für ein menschliches Gehirn. English]
The compassionate brain: how empathy creates intelligence / Gerald Hüther;
translated by Michael H. Kohn.
p.   cm.
Includes index.
ISBN-13: 978-1-59030-330-6 (alk. paper)
ISBN-10: 1-59030-330-X
1. Comparative neurobiology.   2. Brain—Evolution.
3. Brain—Differentiation.   I. Title.
QP356.15.H8813   2006
612.8—dc22
2006000844

CONTENTS

THE
COMPASSIONATE
BRAIN

Preliminary Remarks
and Safety Precautions

Doubtless you drive a car, and you wash your laundry in a washing machine. You use a mobile phone, surf the Internet, shoot your own vacation videos, watch television, and listen to music on CDs. I don't know what other useful or useless gear you may have acquired in the course of your life. But one thing I know for sure: the more complicated and more expensive the equipment was, the more carefully you studied the instruction manual that came with it that described how to use it and how to enjoy it for the longest possible time.

You also possess a brain. And you use it more often than you think. At least much more often than you use all your other gadgets and machines. You use it to get along in life and, at least now and then, you get a little pleasure out of it. But you have, until now, never consulted a user's manual for it. Why not?

Have you simply assumed that your brain is already working properly all by itself? Then unfortunately you have been making a mistake. It only works the way it does due to the circuitry that has been installed in it. And this circuitry—as well as all the further problem-solving circuitry that can be installed in it in the future—depends very much on how you have used your brain so far and what for. So maybe it would have made sense to have checked up sooner on how you were using it. It might conceivably be the case that your usage pattern so far has been leading to a future in which there will be many tasks beyond your brain's capacity.

Or have you simply been taking the approach that, since you have not had to spend a lot of money on this thing and have simply always had it, it needs no further attention on your part? That is also a mistake. There are many other things that just come your way as gifts that are not dead but still living and developing—like children, like relationships, yes, even like your dog or your vegetable garden—all of these need very special attention and painstaking care. And this is true for your brain as well.

Or you may have been hoping that an almighty creator made your brain—or else that your almighty genes have shaped it—once and for all in such a way that it is ideal for taking care of you in this world in the best way possible, and thus there is nothing about it that needs changing. It is nice to think that either He or they (your genes) are responsible for what becomes of your brain instead of you, but this assumption is also a mistake. It is true that every human being's brain is different. Each brain is unique, and from the very beginning each one has had very specific weaknesses and very specific strengths and abilities built into it. But what becomes of this basic makeup of your brain in the course of your life—whether particular weaknesses are compensated for or made even worse, and whether particular abilities are developed or suppressed—depends on how you use your brain and what you use it for.

This may all sound a little disturbing, but sticking your head in the sand won't change it. You'll have to pull your head out of the sand at some point, and admit that your rationalizations are not really solid, but are just plain old excuses. There is only one good reason you can put forward in good conscience for why you haven't yet given any thought to how you use your brain: no one has ever explained to you why you should. That is why I have written this user's manual for you, and I am very happy you have come across it.

I have been working for many years in the field of brain research. Like many other scientists working in this field, I have tried to find out how the brain actually functions. Like all the other researchers, I went as far as I could in cutting up the brains of experimental animals into smaller and smaller component parts and measuring whatever there was to measure about them. I have grown cultures of the various types of brain cells in laboratory dishes and have observed what they developed into and what they were capable of doing. And like so many other brain research scientists, I have done experiments with animals—mostly lab rats—to investigate the effects of various treatments and surgical interventions on their brains.

I still find it fascinating how much there is to dissect, measure, and study in such a brain. But by now I no longer believe that we can succeed in this way in understanding how any brain, to say nothing of a human brain, functions. On the contrary, this kind of research leads us to regard whatever can be especially easily dissected, measured, and studied as being of particular importance in the functioning of the brain. And because researchers are quite happy to talk about that which seems particularly important to them, and because the media are quite delighted to publicize these kinds of new developments, more and more people have gradually come to believe, for example, that happiness results from heightened endorphin secretion, that harmony is produced by plenty of serotonin, and that love comes from particular peptides in the brain. They think that the amygdala is the source of fear, the hippocampus is the source of learning, and the cerebral cortex is the source of thinking. Now in case you have heard of any of this stuff, you can just go ahead and forget about it. The same goes for any claims that particular genetic configurations are responsible for what goes on in your brain. There are no genes for laziness, intelligence, melancholy, addiction, or egotism. What do exist are different basic tendencies, characteristic predispositions, and specific vulnerabilities. But what ultimately becomes of these depends on the conditions for development each of them encounters.

But overvaluing the partial pieces of knowledge that scientists gallop after in the latest high-tech boots is not the only obstacle to understanding what goes on in our brains. Another big problem can be compared to stumbling around in old shoes that have long since ceased to fit. Ideas that appear and for various reasons are regarded during a certain period as extremely accurate, later frequently come to be regarded as dogmas and are presented as such, often by highly respected and admired authorities. Such notions tend to persist for decades. Now if conceptual models describe accurately the realities they refer to, there is no reason to oppose them. But since they only rarely do, most theories in time become like dreadful, ill-fitting shoes that are a major impediment to forward movement.

I myself, like many scientists in the field of brain research, have spent a long time walking around in these kinds of old shoes. The idea that oppressed me the longest and the worst was the dogma that the brain's neuronal circuitry is immutable once it has taken form. This idea was put forward by a pioneer in brain research, Ramón y Cajal. At the beginning of the twentieth century, through the use of new staining techniques, he discovered that the brain is not just a big undifferentiated pudding (a so-called syncytium), but is composed of a countless number of nerve cells that stay in contact with each other through their multi-branched extensions. By means of his stained cross-sections of the brain, he was able to show that this whole thick tangle of nerve-cell extensions gets denser and denser in the course of the brain's development, but that later on, in old age, it begins to thin out to one degree or another. His conception of the brain was adopted by subsequent researchers, and for nearly a century it shaped the thinking of most neurobiologists, psychologists, and psychiatrists. It also became established as a basic conviction in the minds of certain circles in the general public.

In the meantime, it has emerged that the brain remains structurally malleable to a great degree even in adulthood. It is true that after birth, nerve cells can no longer divide (with a few exceptions); however, throughout the lifetime of a human being, the nerve cells of the brain do remain capable of adapting their complex patterns of interconnections to new conditions of use.

The most important and lasting influence in humans on how the neuronal networks and interconnections existing in the brain are used is a thing that is particularly difficult to measure. The best term you can come up with to describe it is *experience*. This refers to the knowledge built up in the memory of an individual about strategies for thinking and behaving that have proved particularly successful or particularly unsuccessful in his life up to the present, that have been confirmed as such again and again, and that, consequently, he now considers particularly appropriate or inappropriate for solving future problems. Such experiences are always the result of a subjective evaluation of his own reactions to perceived changes in the external world that he regards as significant. As results of subjective evaluation, they are distinct from all (passive) experiences and (passively) adopted knowledge and skills to which he has as yet to attribute significance for coping with the problems of his life. Because he becomes embedded in a system of social relations that keeps getting more complex—this normally starts in early childhood and then goes on actively later in life—the most important decisions a human being can make in the course of his life are psychosocial in nature.

It took a long time for me to finally realize that what guides us in all our decisions is not our mind or our consciousness. It is also not the knowledge that we have learned by rote or have adopted from questionable sources. Rather, it is the experiences we have accumulated in our development up to now. The experiences a person has had in the course of his life become firmly anchored in her brain. They define her expectations; they steer her attention in very specific directions; they determine the valuation she puts on what she lives through and how she reacts to her surroundings and what impinges upon her from the outside. Thus in a certain way, these individually acquired experiences are the most important and most valuable treasure a person possesses. She can use it not only for herself but also try—especially if she has ever had the experience that giving brings a great deal of joy—to pass it along to others. The special quality of this treasure trove of experience is that in using it and sharing it, it does not become smaller but gets bigger and bigger.

And if you happen, as I do, to work as a brain researcher in a psychiatric clinic, you not only acquire new experiences, but certain new thoughts and concerns arise in your mind as well. I see patients who are overwhelmed by certain feelings and emotions and have lost the ability to control them. Driven by these emotions, these people sometimes develop ideas that appear to outsiders to be crazy. Many of them feel persecuted or feel that they are being controlled by alien forces. Some have the feeling that they are falling apart or breaking down into different personalities. Some develop feelings of omnipotence and maybe think they are God or Napoleon. Still others feel small and negligible. Others are compulsively driven to be in control of something or other.

On the other hand, I also sometimes see people who are not patients but who are nevertheless driven in a similar way by certain feelings. There are people who consider themselves indispensable and think their opinions are universally valid, and people who have a low opinion of themselves and either prefer to keep their mouths shut or only to repeat what others say. There are people who are dominated by the feeling that they must acquire power and influence and are ready to do anything to attain this goal. And there are those who simply want to be left alone and who are indifferent to practically everything that goes on around them. There are those who simply have to get excited about everything, and those who have a continuous and urgent need for distraction. Many are only capable of keeping at bay the continual flare-ups of malaise and discontent they undergo by means of immoderate eating or with the help of legal or sometimes illegal drugs.

Not only in clinics but everywhere, there are people who behave self-destructively, inconsiderately, egotistically, narcissistically, with indifference, calculatingly, contentiously, self-importantly, and irresponsibly and who thereby cause tremendous damage. In economic terminology this sort of damage is referred to as "friction loss," and economists consider the elimination of friction loss the number one prerequisite for increasing the gross national product of any industrial country. If you ask these people why they behave in such a destructive and selfish manner, most of the time you discover that they have no idea that they do. They just have the feeling that they are behaving the way they have to behave, and that everything they do and think is somehow right for them. This is no more than just a feeling they have.

And that is why I am so inspired by the idea of investigating a little more closely the origins of people's strong feelings and emotions. It is only in the last few years that "emotional intelligence" and the "net of emotions" have become popular and hotly discussed themes. Even psychologists and psychoanalysts are no longer content simply to point out that early experiences play a decisive role in determining later basic behaviors and emotions. Now they too want to know how these experiences are anchored in the brain. They want to know how and under what circumstances it might be possible to "overwrite" these engrams with new experiences, how a feeling, once developed, can be changed and replaced or overlaid by a new one. These questions have brought about a good deal of ferment in the last few years, particularly in the area of brain research. Every scientific discipline passes through certain discrete phases in the course of its development. In each phase it arrives at a particular view of the phenomena it studies. On the basis of the understanding it has arrived at up to that point and the knowledge it has accumulated, it builds a specific theoretical structure. To begin with, this structure is still more or less wobbly. Therefore specific attempts are made to brace it with solid building blocks. It is then further consolidated by means of various organizational measures, and protected as well as can be against the destabilizing influence of undermining ideas and conceptions. But despite all this there is something for which there is no complete defense: the additional knowledge that inevitably arises when investigators do further work on particular questions, think further about the relationships among phenomena, and seek solutions to problems. This new knowledge must somehow be integrated into the old theoretical structure. So long as that can be successfully done, everything is fine, and the edifice remains standing a while longer, even though it gradually takes on a more and more eclectic look, in the form of annexes, gables, turrets, extra wings, and storage areas. At some point, however, the building becomes so hard to get around in (so difficult to have a unified understanding of) and begins to look so bad in the landscape that a drastic reconstruction of the old pieced together heap of theoretical structure becomes inevitable—or even a completely new building becomes necessary. In these phases of upheaval, an old paradigm that everyone had before found completely satisfying is replaced by a new one. The new paradigm offers the possibility of relating to all the knowledge that has been accumulated up to that point as still valid, but it integrates this old knowledge into a new structure that also provides room for new knowledge because it is more comprehensive, more inclusive, and just broader than the old one. These phases of upheaval are the most exciting phases in the development of a scientific discipline— less exciting for those who have made themselves completely comfortable in the old house, but more exciting for all those who felt the old house was too confining, too musty, and just too hard to see their way around in.

The classical natural sciences (astronomy, mathematics, physics, and chemistry) already have this kind of paradigm change behind them. They have all passed through a phase in which they first accumulated, described, and sorted observable phenomena. Then things were all broken down into their component parts, and wherever possible, the properties of these parts were studied as precisely as was possible. After a long enough time had passed in which scientists tried futilely to understand the whole out of their increasingly precise knowledge of the parts, a stage was eventually reached where some of them began to look for invisible forces and dimensions hidden behind objectively observable and measurable phenomena. Names like Copernicus, Kepler, Schrödinger, Einstein, Bohr, Heisenberg, and Planck are associated with these turning points in our understanding of the world on the level of the classical natural sciences. But since most people couldn't care less if the Newtonian laws are only valid when things are neither too small nor too big, if there is such a thing as bent space, if time is only relative, and if waves and particles are interchangeable, these new ways of looking at things have not had that great an effect on our lives and on our way of seeing ourselves.

On the other hand, in biology things are quite different, as they are also in brain research, where the kind of transformation we are talking about is just beginning to happen. Biology is a relatively young discipline in the world of natural science, and its object—life, in all its manifold forms—is so complex that biologists in many areas are still in the gathering, describing, and sorting stage. In many other areas they have already passed to the stage of breaking things down and have begun to gain as precise as possible an understanding of component parts. They have pushed forward to the level of individual molecules, have deciphered the genetic code, and have discovered countless numbers of the signals, signal-transmitting substances, and receptors with whose help information is exchanged within cells, among cells and among organs, and finally even among organisms. They can in part precisely describe how particular life forms have arisen in the course of phylogenetic history, how the information necessary for this was passed down from generation to generation, and how it was used in the elaboration of specific physical characteristics in the course of the development of individuals.

All these are important pieces of knowledge that have contributed significantly to the fact that today we understand the phylogenetic history of human beings better than ever before. We now understand how little human cells differ from the cells of other living beings, how little human organs differ from the organs of other mammals, and how little human modes of behavior differ from those of our relatives in the animal world. That is why Desmond Morris called us "naked apes," forcing us to look at a fact that Darwin pointed out earlier but that we were completely unwilling to admit—that we are no more than a part of nature—and in some ways a very inadequately equipped part. We are not almighty beings, and certainly not the centerpiece of the world. Rather, like all the other creatures, we are embedded in nature as a whole and are dependent upon and intricately bound up with her.

And that is precisely the special point in which the findings and insights of biologists and the brain researchers differ from those of classical natural scientists. Biology and neuroscience not only provide us, as all the other natural sciences do, with ever new and useful knowledge about the world that allows us to shape it in accordance with our ideas. They also continually bring to light new knowledge about us humans, knowledge that helps us make sense of ourselves and of our place in nature.

For a long time I stayed within the old ideational structure adopted from the classical natural sciences—like so many other biologists and neuroscientists—where only one question with regard to the brain was allowed: How is it made and how does it function?

But if the structure and thus also the functioning of our brain depends in a very critical way on how we use it and what we use it for, then does not the crucial question really become: How and for what purposes *should* we use it so that the potentialities built into it really can be fully actualized? In this user's manual for the human brain, I try to answer this question. I base what I have to say on data from the field of brain research that have come to light only in the last few years, data that have made it possible for us, today better than ever before, to understand why and how the way we use our brains makes a difference.

For decades the presumption was that the neuronal pathways and synaptic connections established during the brain's initial development were immutable. Today we know that the brain is capable throughout our lifetimes of adaptively modifying and reorganizing the connective pathways that it has laid down, and that the development and consolidation of these pathways depends in quite a major way on how we use our brain and what for.

A few years ago, no researcher in the field of brain science could have conceived of the possibility that what we experience could be capable of changing the structure of our brain in any way. Today most scientists who study the brain are convinced that the experiences of our lives do become structurally anchored in the brain.

Until quite recently, it was held to be self-evident that human beings have a big brain to make it possible for them to think. However, the research results of the last years have made it clear that the structure and function of the human brain is especially optimized for tasks that we would subsume under the heading of "psychosocial competence." Our brain is thus much more a social organ than it is a thinking organ.

As recently as a few years ago, everything that had anything to do with feelings and emotions appeared suspect to brain researchers. But lately they have begun to understand how important feelings and emotions are, not only in orienting perceptual and thinking processes, but also for the way in which early experiences become anchored in the brain. Thus the great role feelings play in determining our later basic attitudes and convictions has become clear.

For an entire century, it was hotly debated whether the thinking, feeling, and behavior of human beings was determined more by inborn behavioral programs or by experiences acquired in the course of our lives. Nowadays advocates of psychological and psychosocial determinism are beginning to accept the view that human feeling, thinking, and behavior have a material, that is to say, a neurobiological basis. On the other side, believers in the biological determinism of psychological phenomena have come to acknowledge that, at least in humans, psychological processing of social experiences is of considerable importance, both for the stabilization of particular genetic inherencies within a population and for the formation of particular neuronal and synaptic connective patterns in the brain.

A lot of this new data, that has appeared in a flood of scientific publications in the field of brain research, went largely unnoticed by its potential users, such as doctors, therapists, and educators. It did not receive major play in the media, and it will be years before it begins to appear in school textbooks. The fact that so many people understand either nothing or very little of what is going on in their heads and the heads of their fellow human beings makes both the writing and reading of a user's manual for the human brain a difficult undertaking, and one that is not without dangers. I have made an effort to write this text in such a way that the most complicated

part is found at the beginning, right here in these preliminary re-
marks. If you have gotten this far, then the rest is child's play.

But watch out: this child's play could get serious fast. If that
happens, it might happen that nothing will end up the way it was
before. Including your brain.

Overview

This user's manual is not for people who are afraid of change. In the course of their lives many people put on blinders and tinted glasses of varying thicknesses and colors. In this way they can often go a fairly long time without noticing that something is going on around them that is really compelling them to change. These mental tinted glasses and emotional blinders are made up of safety precautions and defense mechanisms that are necessary at certain times. But they must be removed when a person really wants to make free use of his brain. Therefore it is urgently required that you take these things off at least while you are reading this manual. Since this not only goes against our habitual patterns but is also uncomfortable at first, we will begin by describing how to free one's brain, at least for a time, of everything that gets in the way of an unobstructed, impartial, and clear way of seeing. By the time you finish reading the rest of the chapters, you will definitely have lost the inclination to walk around with your old glasses and blinders on, at least voluntarily.

What a brain can be used for inevitably depends on how it is structured. And how a brain is structured, in turn, depends on what it has been used for up to now, indeed not only by its present owner, but also by his or her ancestors. These ancestors have tested a variety of blueprints and building plans (in the form of specific genetic configurations), and when these functioned to any degree, they passed them on to their offspring. But a genetic blueprint of this sort is still far from a finished brain. In order for it to be fleshed out once again into a competently functioning brain, the parent generation must pass on to their offspring not only the blueprint but also the materials needed to complete the actual building job. For simple program-guided structures, like the ones worms, snails, and insects creep around with, this does not amount to much. The eggs, which contain the necessary building materials, just have to be laid in a place that is suitable for the offspring's development. For the most part, the rest happens all by itself.

Most vertebrates already have initially programmable structures—in this case, brains—that can be shaped at the beginning to a certain degree by their own experiences. In such cases, during the phase of brain development, parents have to create and maintain conditions in which the offspring can learn what they need to know for their later lives. That is already more difficult than in the case of snails and worms, but normally it goes off without a hitch as long as the world in which the offspring grow up does not change too much.

This last point also holds true for cases where the genetic blue-prints enable the formation of a brain capable of learning and con-tinuing to be programmed throughout an entire lifetime. Only human beings have a brain like this; we have had it for about a hundred thousand years. So for about four thousand consecutive generations not very much has changed in the ability of our genetic base to produce a brain the more refined level of whose structure depends throughout its entire lifetime on how we use it and what we use it for. Thus the individuals of all of those generations had to learn anew and afresh, within the life conditions created by their parents and their ancestors, what life required of them. Sometimes that was a lot, sometimes not that much. There have been times and places in which humans have been able to produce and main-tain favorable conditions for the elaboration of highly complex and finely networked brains over a number of generations. But there have always also been times and places in which conditions did not allow the genetic potential for developing and training highly complex and networked brains to be so fully utilized. In this re-spect, not so much has changed today. Today there are still people who are lucky enough to grow up in a world where they have the opportunity to maximize their genetic potential for a brain that is capable of learning their whole lives long; and there are others who are obliged to find simpler solutions to ensure their own survival and that of their offspring.

How different brains can actually be from each other, particu-larly those of us humans, why they are so different, and to what a great degree the neuronal interconnections established in them determines the further potential for their use, will be discussed in the second chapter. There we will also give a clear picture of the characteristic that distinguishes a human brain from all program-run structures: the capacity throughout our entire lives to break down interconnective neuronal circuits already established in the brain, repattern and restructure them, and with them the mental and behavioral patterns that they determine—including even apparently unshakeable basic convictions and emotional patterns. It is because of this capacity that the human brain alone is capable of erasing and overwriting already existing programs as soon as they begin to hamper the further development of our mental and emotional potentialities.

Like all learning-capable brains, the human brain is also most deeply and enduringly programmable during the phase when the brain is developing. Thus the most important installations in your brain were in place long before you were able to read this user's manual. Important experiences acquired in early childhood and youth led to the stabilization of certain neuronal pathways. These connective patterns, once they are facilitated and broken in, become especially easy for our perceptions and experiences to activate, even later in life. They thus become determinative for "what goes on inside us," for how we feel, think, and behave in certain situations.

In order to break down this kind of programming later on, already existing installations have to be rendered conscious and acknowledged. This is what we will talk about in the third chapter. There we will first describe what developmental conditions are necessary to make optimal use of our genetic potential for the elaboration of a highly complex, densely networked, learning-capable brain. Since only a small proportion of human beings have the good fortune in their childhood to actually find themselves in such conditions, we will also concern ourselves in that chapter with the traces left behind in the brain by less ideal or outright inadequate developmental conditions.

A brain that is capable of learning throughout life is also changeable throughout life. This means that installations that were put in place during the phase of brain development that were one-sided, unbalanced, or defective are to a certain extent correctable during the adult phase. How to make such corrections is the subject matter of the fourth chapter. The goal of all the corrective measures is to restore a lost inner balance. Often in the course of our development, imbalances in the relationship between openness and self-differentiation arise. Dependency on relationships with "important others" can be overdeveloped just as our striving toward autonomy can. Feeling and thinking are then seldom in harmony and all too easily come into conflict with each other.

All of these unbalanced patterns tend to stabilize—mostly over years and sometimes over decades—and end up having a major effect on conditions of brain use. To be precise, they shrink our brain-use potential. Since under these circumstances, further development is blocked, correcting these defective neuronal installations becomes a main prerequisite for the ongoing elaboration of a highly complex, densely networked, lifelong-changeable brain.

Nothing in the brain stays the way it is if the brain does not continue to be used in the same way. And no aspect of the brain can keep on developing and becoming increasingly complex if it has no new tasks to accomplish and no new challenges to meet. These two core propositions summarize what we describe in the fifth chapter under the heading of "Maintenance and Servicing." If the human brain were no more than a complicated thinking and memorizing organ, then the best way to maintain and service it would be by playing intellectual guessing games and learning the phone book by heart. If it were no more than a central coordination organ for directing vital bodily functions and complex sequences of movement, it would have to be trained and stimulated by means of physical-conditioning programs and other bodily exercises. If the primary function of the brain were to process our perceptions of the external world and our inner body world and translate them either into unspecific images, feelings, and dreams or into specific reactions, then the essential would be to educate and train it further in this ability to perceive and process perceptions. And if the main reason we possessed this big, learning-capable brain of ours were to compete with other people, then it would be advisable to find better and better strategies for outdoing them, cheating and deceiving them, subjecting them to our will, and otherwise exploiting them for our own purposes.

Even though in the past it has repeatedly seemed, and often been propounded, that the special purpose of the brain was to be used in just one of these ways, our up-to-date knowledge today makes it clear beyond a doubt that all of these purposes are equally in play. Thus the master stroke in using our brains properly is to keep creating conditions that not only make it possible, but also indispensable, to use all these capacities of our brain simultaneously to the greatest extent possible and to train and develop them equally. What conditions these are, where we can find them, and how we can use them to maintain and service this human brain of ours, will be explained in detail in the fifth chapter, the main part of this manual.

Anything that is complicatedly structured also reacts very sensitively to any kind of disturbance. The brain is the most complicated organ we possess. The fact that the process of developing a human brain can happen successfully and can continue despite its enormous vulnerability to disruptions borders on a miracle. The probabilities incline far more in the direction of the human brain being blocked in the full realization of its potential by conditions unfavorable to its development and continued high-level use. The sixth and final chapter deals with the major kinds of breakdowns that we can normally expect to occur frequently. The most common cause of breakdowns is serious abuse. Such abuse usually occurs quite early on, during the time when it is mainly our parents and other persons close to us in the early stages who are establishing how and for what purposes we use our brains. Later the circle widens to include other people who influence the way we use our brains. From these other people we adopt ideas and conceptions that seem particularly well suited to help us cope with the world we are growing up in. Thus the way we use our brain depends not only on the challenges we have to face in dealing with the world around us, but also on what ideas are provided by other people that we can adopt for dealing with these challenges. The world that most people grow up in is a world that is more or less consciously structured in accordance with the standards of previous generations. That world is not necessarily a particularly human world and therefore not necessarily a world in which optimal conditions for the development of a human brain prevail. The less fully human conditions are met, the more members of new generations are forced to go against what should be the proper operating instructions for the use of their brains. Under such circumstances, the miracle of a fully developed human brain becomes rarer and rarer, and over time, the thing that happens the most—in this case, impairment of the full development of the brain—comes to be seen as the normal thing. And when this point has been reached, we are left with only three possibilities: (1) to doubt the omnipotence of our creator, (2) to alter the basic genetic setup of the brain in such a way that the brains produced by it are better suited to existing circumstances, or (3) to alter existing circumstances in such a way as to make possible the development of brains that are increasingly more human in nature. The first of these possibilities has already been pretty well exhausted; the second we continue to try out from time to time. The uncomfortable third possibility is something we keep trying to postpone.

REMOVING THE PACKING AND PROTECTIVE MATERIALS

*E*verything alive, including a brain, finds itself in a big dilemma: It must be sufficiently open to take in everything it needs for the growth and maintenance of its inner order, but at the same time, it must be sufficiently closed to prevent external disturbances from penetrating into its inner world and threatening the stability of the inner order that has been built up there. The brain has solved this dilemma in a particularly clever way. It can open its apertures onto the external world particularly wide when it needs to pay close attention to whether or not something threatening is going on out there, and it can simply close them up when what is going on outside does not appear particularly threatening. In case things do become really dangerous, it always has the option, with the help of the legs or maybe wings of its owner, to make a getaway at top speed, or with the help of its owner's teeth and claws, to defend itself against an attack on its inner order.

Many brains are very sensitive and have the ability to perceive threatening changes in the world outside while these changes are still in their very early stages and have not yet really materialized. These brains can assess in a foresighted manner whatever is confronting them and thus take defensive action sooner and more effectively against threats to their inner order. They can recognize and ward off dangers while they are still some distance off. They can seek out solutions before it is too late. But since in the long run, this approach is a great drain on one's energy and can lead to an excess of predictive thinking that results in a total muddle in one's head—to the point where you even end up thinking you hear the grass growing—the brain only very seldom makes use of this extraordinary anticipatory ability. Rather than continually exercising such extraordinary alertness, we prefer to sit and daydream, do the occasional crossword puzzle, have a constant stream of music and colorful images going in the background, and just trust that everything will turn out all right. The longer a brain is used in this latter fashion, however, the harder it is for it later to rouse itself when something important is actually going on.

For you to be able to get somewhere with this user's manual, you are going to have to extract your brain from the protective padding in which it is normally so nicely and comfortably packed. You might as well know that from this point on the comfort you have hitherto been enjoying is over once and for all. Any time a brain is suddenly taken out of its wrapping and stirred up, a certain turmoil begins to take over. This is quite unpleasant, because it also extends into our body. Our heart begins to race, we get butterflies in our stomach, we begin to perspire, and we might even have to go to the bathroom. A stress reaction is what this thing is called, and the feeling that goes along with it is called fear. When you experience this, your brain has really woken up. It has opened all its sensory receivers and channels wide and is now trying to find out where the disturbance to its inner balance is coming from and how it can be dealt with.

But as soon as it figures out that in reality no more is actually going on than that you are reading a book, your brain will immediately want to calm down again and settle back into its bed of protective padding. You can only prevent this by explaining to it that this book is about how to operate a human brain properly. Then it will surely rally.

But now the difficulties are just about to really start. Because when you read the operating instructions for a *human* brain, your brain immediately will begin to fear that the comfortable mode of operation it has been enjoying so far is now over. So it will start to say things that it hopes will help it get back to its old peace and quiet. It will deploy protective devices in the form of all kinds of defensive statements.

First it will say that the world is full of books that are utter nonsense, and therefore it is hardly likely there is anything to be gained by reading this book. And when you have explained to it that there are exceptions and it should at least wait and give this book a chance, its next claim will be that a brain is far too complicated a thing for us to be able to understand how to use it. To this effect it will cite a whole array of experts who have repeatedly shown how hard it is—even quite impossible—to find out with a brain how a brain works. Then when you have pointed out to your brain that it is not how it works that you want to know but how to use it, it will very skillfully drop the innuendo that what is said in this particular user's manual is probably wrong. Then you have no choice left but to encourage it just to wait and see. Try to do this in a loving and gentle manner. Your brain is more timid and fearful than you think. Perhaps you will win it over if you explain to it that surely many other people will also be reading this set of operating instructions.

2

OPTIONS FOR ASSEMBLY AND POSSIBLE APPLICATIONS

A *stationary collection of cells does not need a brain. It cannot go* over there where things are better, nor can it run away from here when things get nasty. For it, a brain would be a sheer luxury, something it could not begin to do anything with. The brain would atrophy and just drop away at some point without the loss even being noticed.

This is what happened to tapeworms. Tapeworms' ancestors were once quite mobile worms. They had a nervous system that coordinated the contractions of their many muscles in such a way that the whole worm was capable of forward motion. This nervous system had the ability to process incoming signals from its sensory receivers, and this allowed the worm to creep to places where there were no dangers, where there was something to eat, and even under certain circumstances, to a place where there was a suitable partner, ready for mating. Later on, at a time when even bigger and more complicated animals than themselves had arisen, some of these worms succeeded with the help of their primitive brains in locating a particularly pleasant environment—the intestines of these larger animals. There they found plenty of food, and there were no threats to deal with, as long as the host survived. Little by little, the worms lost their mobility. On the outside of their heads, there developed a crown of hooks that enabled them to hold on tight, and inside their heads, everything disappeared that was no longer of use in this worm paradise. Without noticing it at all, they had lost a brain—which was not that big in the first place—and not only that, but they also soon lost the ability to develop one.

Up to the present time, the same thing that happened to tapeworms has also happened to all other parasites. To begin with they used their brains in a particularly cunning fashion to create a comfortable lifestyle for themselves, and when they finally achieved this, they began to turn into zombies. So one possible application for the brain is to use it to acquire a habitat in which a brain will no longer be necessary.

The same holds true for assembly options—that is, for the structuring of the brain and its various regions. Building up an individual region of the brain that is responsible for certain functions depends on what one needs a brain for and thus what one ends up using it for. Let's take the mole, for example. The mole's ancestors were insect eaters and therefore had be able to see and jump around at least fairly well. But since under these circumstances, they continually ran the risk of being eaten by bigger animals, it was definitely beneficial for them from time to time simply to bury themselves. If it turned out that they also found enough to eat under the ground, these primordial moles soon no longer had any reasonable motive for ever coming to the surface again. They dug their tunnels and caught their earthworms or whatever else was down there. There was nothing to look at in these dark tunnels, but they had to be able to smell and hear well. The moles who most rarely had a yen for light and sunshine and who developed the biggest front shoveling gear must have been the ones who lived the longest and left behind the most offspring. Pretty soon all the offspring were as blind as the older moles, had long noses, and such big digging attachments that none of them could jump around anymore. Then their visual cortex, having become as useless as it had for them, began to atrophy. By contrast, the regions of their brains that were used for smelling and hearing gradually became more extended and better developed.

This is the fate of all specialists. Initially they use all their senses and their whole brain to find a niche where the living is fairly easy. And once they have found it, from generation to generation their brain and their entire body structure adapts itself better and better to the conditions prevailing there. The more one-sided these conditions are and the more successful the adaptation process is, the harder it becomes for these specialists later ever to get out of their niche.

So a second option for brain application consists in acquiring a habitat in which particular parts of the brain are very much used and—at the expense of other, less needed regions—more and more highly developed.

The third application possibility is the most interesting one, but also the most difficult. The brain can also be used to acquire a habitat that makes such complex demands on the brain's capacities that all its abilities are called upon and developed to the same degree. In the course of evolution, the only life forms that brought off this master stroke were ones that for some reason or another did not do so well at occupying and defending a habitat where the essential thing was to be able to see, hear, or smell particularly well. Or to run, climb, swim, or fly particularly well. Those who lost the race in all these specific disciplines, who were able to do a little of everything but were not especially good at anything, these apparent losers in the evolutionary process, had as their only remaining chance the option of holding open the further development of their brain for the greatest possible number of uses. They did not need a brain that at the time of birth was already so mature that it permitted them to fit into a very definite, highly specific habitat as quickly and perfectly as possible. A very strict genetic program of the type that steered the brain development of the offspring of all the specialists in a very definite direction was of no particular use for these jacks-of-all trades-but-masters-of-none. In the big race for the best survival niches, they had in some sense missed the starting shot. The specialists were long since off and running, and as far as the latecomers were concerned, the race was pretty much over. To follow along behind was senseless. The only thing they could do was to try to stay the way they were and stick it out until the others— with their strategy of cultivating specific achievements and abilities—either became exhausted or somehow got stuck.

The way things actually happened, there were soon no more habitats on the earth that had not been taken over and occupied by some specialist or other. In the water, on the land, and in the air, the most varied life forms spread and multiplied, and the world became more colorful and more full of forms and voices than it had ever been. And at the same time the world began to change at a rate that had never been known before. The two of these things together—the increasing complexity of the external world and the increasing dynamism of the changes occurring in this world— allowed the laggards the chance to pass up the specialists without catching up with them. Now for the first time, in this multifarious world that was growing ever more changeable, it began to be advantageous to have a brain with which one could both smell and see, and also hear, feel, swim, and climb, and even in some cases fly.

And so what had to happen, happened. The specialists played out their expertise and came to the end of their developmental potential. By contrast, the generalists, who could do a little of everything but were not particularly good at anything, and who up to now had been able to hold their own in the world against the specialists only with a great deal of trouble and effort, now really got going. There was an incredible amount to discover in this world created by so many hearing-specialist artists, sight-specialist artists, smell-specialist artists, and all kinds of other specialist artists—if you could use your ears, eyes, nose, and skin equally well and synthesize everything you heard, saw, smelled, and felt into the most complete possible picture. With such a talent it was possible to grasp complex changes in the external world on several sensory levels simultaneously and use this information for foresighted and circumspect reactions. And of course this all worked better the less the neuronal pathways operative in the brain had been predetermined from the beginning by genetic programs. In this way, from originally strictly program-guided structures, there gradually arose increasingly open pathways of neuronal connectivity that were no longer exclusively genetically controlled. In such cases the final and definitive pattern of the neuronal pathways was stabilized only later on, within the context of individually encountered conditions of use. Structures that were entirely program controlled turned into structures that were subject to initial programming and later on even into structures that were programmable throughout an individual's lifetime. At the end of this developmental movement, there finally arose a brain that was capable of establishing the terms of its own usage and thus to a certain extent of structuring itself. It could decide for itself what was to become of it. With such a brain, our ancestors set out to make a world in which they could create the conditions for the use of their brains in accordance with their own ideas. How often in the course of doing this, they strayed off onto errant paths can be read about in our history books.

2.1 Program-Controlled Structures: Brains of Worms, Snails, and Insects

The first and most primitive of all nervous systems were developed by animals who through accidental program changes in their genetic material succeeded in forming something like a hollow cone out of a raw heap of similar cells. In the liquid-filled hollow space inside this cone arose an inner world of its own, which was shielded to a great extent from the disruptive influences of the alien outer world. Of course such a cone of cells could only stay alive as long as it was able to maintain the conditions prevailing within it when the outer world began to change in a threatening manner. This could only occur if all the cells were constantly informed as to whether outside in the world something important was happening and whether inside, in the inner world, everything was still in order.

There was only one solution to this problem, and some time or another this solution was found through some accidental alteration of the creature's genetic program. Some of the outer skin cells became retarded in their development and began to wander back and forth in the space between the outer and inner skin, forming extensions through which they remained in contact with one another, with the outer and inner worlds, and with the cells of the inner and outer skin. Thus the outside and the inside became connected, and the whole organism could now, in a concerted action of its cells, react to anything that threatened its inner order. This very first function of the very first nervous system has remained up to the present time the most important task that a nervous system, also our nervous system, has to perform—the maintenance of our inner order.

To begin with, this did not take very much—in fact, the structural outlines of this first kind of nervous system remind us very much of the regulatory system of an air conditioner that keeps the temperature, humidity, and fresh-air supply in a house constant whether it is freezing cold and snowing or swelteringly hot outside. For this purpose, all that is needed is a simple program-controlled structure with the right kind of sensors, which sets a rebalancing regulatory mechanism in motion whenever the values the sensors measure deviate from the ones the unit is set to maintain. Nervous systems set up in this way were completely adequate for life in a world that was not changing too much and in which nothing new and dangerous was happening. Many little cones are still doing quite well in the world they live in today. Many of them, like polyps and jellyfish, have developed an amazing variety of forms.

Most of these forms, however, have gone under in the course of millions of years, because disturbances always arose in their outer worlds that they did not recognize with their primitive nervous systems and thus could not react to efficiently enough. But there were always a few who, because of accidental changes in their genetic programs, were a little different and could perceive certain threats better or sooner and could therefore react differently and more efficiently to them than their fellows of the same kind. If they had a paddling apparatus or contractile cells with whose help they could move out of danger zones and toward sources of food, for example. This locomotive apparatus, however, could only be used effectively if there was also a nervous system that coordinated its movement in such a way that the entire organism could be steered in a particular direction by it. And once this had come into existence, not only did the originally cone-shaped creature take on an ever more streamlined—and thus more wormlike—form, but also whatever was happening in front of them now became definitely more important than anything that was happening behind. For that reason, the very first accumulation of nerve cells developed in the place where all animals still have their brains: in front, in the head. And the more sensors that could be concentrated there for the perception of physical (tactile, optical, auditory) or chemical (gustatory, olfactory) changes in the outer world, the better, the surer, and the sooner threats to the inner order could be recognized and warded off; and not only that, but the better, surer, and sooner places could be spotted in the organism's environment where especially favorable conditions for the maintenance of its inner order prevailed. And all of this functioned the more effectively the better the pieces of information coming into the various sensors—that is, the different sense organs—could be bound together into one general impression of the changes taking place in the outer world and the conditions prevailing in the inner world. And that is what went on in this accumulation of nerve cells in the head. The bigger it was and the more neuronal pathways and cross-connections it had to the various sensory receptors, the more complex this primitive brain became and the more precisely the picture it composed of perceived changes in the external world corresponded to what was actually going on out there in reality.

Whatever stage of complexity these brains reached in their path of development passing through worms, snails, and insects, they all remained strictly dependent on their existing genetic programs and the structures they controlled. The genetic base required for this came into existence over inconceivably long periods of time, initially through accidental alterations in the DNA sequences that were already present—in fact, really through the mistakes that repeatedly occurred in the transmission of genetic programs from one generation to the next. As though through the action of a generator of random events, there continually arose new, supplemental, redundant, disjointed, and otherwise newly arranged sections of DNA, which resulted in more or less significant program changes. As a result of the fusion of parental cell nuclei in sexual reproduction, these genetic configurations kept getting further blended and mixed. An individual who had received a genetic program that made it possible for him to build up and maintain a certain inner order remained alive and produced offspring. The genetic programs of those who left behind the most offspring proliferated and spread. The others, as unsuitable prototypes, got left behind. That was nearly all of them, because the theoretically unlimited possibilities for genetic program alteration in practical fact always had very strict limitations to face. Only a very small proportion of everything that was possible turned out to be actually livable. Changes could not be too big or too drastic. They had to fit in with what was already there, and they had to not get in the way of what was already functioning. That was particularly the case regarding the structural plans for the brain. Above all, changes had to be conducive to the production of more offspring or else of offspring more capable of surviving. The number of offspring could be raised by a whole array of the most different kinds of program changes. But their ability to survive could only be improved in one way: through a better-functioning brain. So from the ever-broadening range of genetic variability, those programs were repeatedly selected that were suited to the structuring of a brain that could more quickly perceive and more effectively respond to changes in the outer and inner worlds.

Since as a result of this kind of broadening of the range of genetic programming, the inner world of the life forms that were developing inevitably became increasingly complex, and since their outer world was increasingly being shaped by the activities of other life forms that had never existed before, the neuronal circuitry and associative networks being formed in the brains of these first invertebrate animals had to become more and more complex. That made possible improved control of vital bodily functions (metabolism, circulation, respiration, digestion, etc.), of behaviors important for survival (attack, defense, flight, acquisition of food, sleep), and of species-specific reproductive strategies (partner seeking, partner selection, mating, care of offspring).

From the very beginning, the invertebrate animals were dependent on reproducing on a massive scale in order to give rise to a sufficient number of genetic variants. Out of this pool, through natural selection those variants were favored that fit into the particular species' habitat better than any of the others, because they could do something essential in that particular habitat better than any of the others could. Like a series of robot models developing toward an ever more efficient solution of particular tasks, there finally arose from the primitive nervous system of the coelenterates, worms, and snails such complex things as the brains of arthropods, which functioned in every conceivable kind of habitat—in the water, on land, and in the air. In so doing, some of them developed highly complicated species-specific behaviors. With bees, termites, ants, and other colony-building insects, there even arose collective, labor-sharing communities. Every mode of perception, food-acquisition, defense, or locomotion that could possibly be used for survival and every behavioral strategy that could possibly be used in partner seeking, mating, or the safeguarding of offspring probably made an appearance in the long course of development from worms to insects, taking the form of specific genetically programmed neuronal pathways. And the ones that worked were picked out by the selection process.

By the time the end of this long road had been reached, it was the various specialists who had conquered the world and divided it among themselves in the form of a very wide variety of ecological niches. The better these specialists were adapted to the conditions prevailing in their niches, the more they flourished in them. However, this lasted only so long as everything in those niches remained largely as it had been. But when the habitat they had occupied began to change, with their specialized brains, they soon came to the end of their possibilities. Such changes came about inevitably, as a result of their own activities, like when they reproduced too fast, or if they were too successful in spreading throughout their inherited habitat. Or through the activities of other species, through gradual shifts in climate, or as a result of sudden catastrophes. In all these cases, the path of developing specific, strictly genetically programmed neuronal connective patterns, hitherto so successful, now led into a fatal dead end. At the point where different, completely new abilities were suddenly necessary, the specialists' expertise proved entirely futile.

Through many small program changes, genetically programmed installations for the control of specific behaviors can be gradually improved. In this way they get better and better at dealing with the tasks they are optimized for. However, it is all but impossible later on to gradually undo special programs that have been produced in this way or to change them to make them optimally useful for other applications. These strictly genetically programmed brains are different in this respect from computers that have been programmed in a particular way. You cannot do anything with them but what they have been programmed for. If the owner of a computer is lucky, he might find a place for his old machine in the museum, and he can buy himself a new one. If the owners of such dedicatedly set up brains were lucky, they ran into a habitat that changed so little that they have still survived in it up to the present day and are continuing to multiply. All the rest have died out. New multipurpose programs could be developed only by those kinds of brains that from the very beginning were constructed in a different way from theirs.

2.2 *Initially Programmable Structures: Brains of Birds, Marsupials, and Mammals*

Worms, snails, and insects belong to the large group of so-called Protostomia (first mouth). In these organisms, the first opening that appears during the early development of the embryo also remains the mouth opening later on. Another group of animals, the Deuterostomia (second mouth), to which the sea cucumber, all vertebrates, and we ourselves belong, develops what is later its mouth from another, second opening, which the embryo forms during the blastocyst stage. In our case, it is not only our mouth but also a number of other things, which in the course of embryonic development develop out of the so-called blastema, that become fixed in form only later and by no means as strictly as with the older Protostomic creatures. This is particularly the case for the nervous system.

In the embryonic stage of these Deuterostomia, more easily than with the embryos of insects, for example, things that develop later from a particular cell group can be thrown into disarray from the outside by means of relatively simple manipulations. The embryonic cells do not "know" for a long time what they are supposed to develop into. This only becomes established through a set of conditions they find their way into in the course of division inside the embryo. All genetic programming does is make each of these cells able to turn into something definite at the point where the conditions it is growing up in begin to change in a very specific way.

Thus the various embryonic cells are in reality controlled not by a single gene but by the set of conditions that arises inside the embryo, which they themselves at the same time take part in determining. We should imagine this set of conditions as a cocktail of all kinds of growth factors, signal substances, hormones, and transmitters. In each area of the embryo and at every point in time in its development, a very characteristic combination of these substances prevails, which causes the cells in question to call up certain very specific genetically stored programs and switch off others. This signal cocktail can be—as it is with the Protostomia—enormously strict and unequivocal and therefore almost impossible to influence from the outside. But it can also be—as with the Deuterostomia— less strict and more ambivalent and therefore easier to change. When that is the case, the conditions prevailing in the external world, outside the embryo, can become important for what goes on inside. Thus embryonic development comes to a certain degree under the influence of certain factors in the external world. This was precisely the structural advantage enjoyed by the ancestors of today's vertebrates. The expression, that is, the implementation and realization of their genetic programs, became open to influence from changes in their external conditions. These changes could have the strongest effect on the system in the organism that developed the slowest and whose development was the most strongly regulated by complex regional changes in the production and secretion of signal substances. That system was the brain. Nevertheless, there was still a long way to go before, from the first Deuterostomic creatures, the first vertebrates arose, with brains whose final neuronal connective pattern was determined by the outer conditions that were present during their early development.

These ancestors of today's vertebrates came into being later than the ancestors of the worms, snails, and insects. They were forms that were still to a great extent undifferentiated, and they remained for a long time at a primitive level of development. Their nervous system was very simply structured. Specific-performance capacity had not yet developed much, and the genetic programs dictating the set of conditions responsible for the further elaboration of the nervous system were correspondingly simple—not very precise and above all not very strict and binding. Given that these ancestors of the vertebrates looked a lot like our present-day sea cucumbers and lived in a similar fashion, a very precisely functioning nervous system was not all that important for their survival. Their nervous system developed slowly and was from the very beginning directed more toward keeping the inner world of these animals constant than toward influencing their outer world through specific behavioral reactions. Their habitat was the sea or the sea floor. And that is where they had to stay for the time being, because the fact that the development of their offspring was highly subject to the influence of changes in outer conditions left them no choice but to lay their eggs where the same conditions always prevailed. In a way, they used the sea as an immense uterus.

Even when the first vertebrates, in the form of lobe-finned bony fish (coelacanths), came up onto the land and later on, in the form of amphibians, began to creep about on it, they remained dependent on being able to find sufficient amounts of warm, salty water as a safe place with constant conditions to lay their eggs in. They could only conquer the land, and later even the air, by laying disproportionately large eggs that contained within them everything necessary for the undisturbed development of their offspring: nutrients, salt, and enough of that still-required substance, water. They had to lay these eggs, which were protected by thick shells, in a place where it was, and remained, warm and moist enough and where there were as few disturbances as possible. This result was best achieved when one of the parents, most often the mother, just immediately sat down on the eggs herself. The mother frogs and reptiles could not do this, because being cold-blooded, at night they were rather cold themselves. There were no warm-blooded animals yet. But the ability to keep the conditions of the inner world constant was already quite widespread. Some of the ancestors of today's warm-blooded animals succeeded, through a number of small changes in their genetic programs, in developing this ability further—to the point where they were able to keep their body temperature constant even when it was not very warm out, and even when it was quite cold. These first warm-blooded creatures were the ancestors of today's birds, marsupials, and mammals. With this ability, they opened up for themselves a world that had been ruled up to that point solely by specialists and cold-blooded animals. These latter creatures, if they did not live entirely in the water, quite quickly became torpid once the sun ceased to shine quite directly down on them. Now, with this one development of warm-bloodedness, at least three doors that had hitherto been closed opened at the same time.

One of these doors led to the shadier and cooler areas of this earth, which up until now had seemed to all other animals to be too chilly and hostile to life due to their own inadequate bodily warmth. The fact that at this same time a climatic catastrophe occurred that led to a long period of cold was just another contributing factor. The dinosaurs and many other specialists were wiped out, leaving the habitats hitherto occupied by them free for the ancestors of birds, marsupials, and mammals.

The second door led into the night. In a world that had been ruled up until then purely by cold-blooded specialists, who for the most part went stiff as soon as the cold night fell, the warm-blooded creatures were capable of turning the night into day. Because of this, they now penetrated into habitats that had in fact long been occupied, but only during the day. In order to cope with night conditions, more and more sensitive senses were needed. You not only had to be able to see better, but also feel, hear, and smell better. And you were best able to cope if you could do all of these equally well. So there was an unprecedented pressure of natural selection on these first warm-blooded animals that drove them as never before to develop sensors for the perception of the most various sensory modalities as well as to develop the neuronal connectivity needed to process these multiple sensory inputs. If you are going to process a great many inputs at the same time, you must be able to associate the data coming in from the various sense organs into a composite picture. The mental picture that thus arises is produced in the brain through a characteristic combination of what has been heard, seen, smelled, and felt, and the significance of this picture is arrived at by comparing it with data that has already been stored previously. The less these perceptual and associative processes run along already established, already fixed neuronal pathways, the better they work. Neuronal installations that are strictly genetically controlled and conduct certain kinds of perceptions along specific neuronal pathways and go on to trigger definite programmed reactions are not appropriate for this kind of complex processing. Nevertheless they remain advantageous when the fastest possible reaction is needed in order to ward off a threat in the most effective possible manner. Therefore two things were needed. The first ancestors of today's warm-blooded animals, when they wanted to use all their senses, needed a brain whose neuronal circuitry was determined as little as possible by rigid genetic programs in their final form. But in order to be able to react quickly and efficiently to dangers and threats, they at the same time needed a brain whose neuronal circuits functioned at the highest possible level of efficiency—that is, circuits that were as strictly genetically determined as possible. For this problem, there was only one solution, and it was only a matter of time until it was found: In addition to the parts of the brain that were responsible for the maintenance of the inner order and for warding off threats and whose setup was strictly determined by genetic programs, there arose a new brain region whose neuronal pathways were not yet fixed at the time of birth, but were only definitively formed, stabilized, and established in accordance with the conditions prevailing for their use, that is, in accordance with the experiences encountered during the early phases of the brain's development. In this new region of the brain, now for the first time, individually acquired, complex perceptions and experiences could be anchored in the form of characteristic patterns of neuronal connectivity.

What kind of experiences these primarily were becomes immediately clear when we take a look through the third door that began to open with the invention of warm-bloodedness. The warm-blooded animals could now hatch and care for their own eggs, either as in the case of birds, outside in a nest; or in the case of marsupials, first in the uterus and then in a belly pouch; or finally, as with mammals, first in the mother's body and later at the breast. In this way, they became independent of particular places, particular times, and particular regions. They no longer had to keep going back, as turtles still must today, to the same places to lay their eggs in order to achieve the necessary conditions for the disturbance-free development of their young. But on top of that—and this was much more crucial—they were in a position for the first time to create by themselves the conditions needed for the raising of their young. This provided an opportunity to take advantage of the reduced level of genetic control of the offspring's development in a particular way. Now at last that which had always proven to be a disadvantage—namely the openness of the offspring's developmental process to external influences and thus the vulnerability of that process to disruption—now became a decided advantage that could be exploited during the offspring's period of brain development. Their brain development was to a certain extent controllable through the conditions that their parents were able to create, since the neuronal circuitry that was laid out in their brains in a basic form yet not fully mature was susceptible to being shaped and programmed by their own early experiences.

In birds, marsupials, and animals, we find many examples of such initial programming that look like genetically determined, inborn behaviors, but when looked at more closely reveal themselves as cases of imprinting occurring early in life.

Colony-building seabirds, for instance, are so strongly marked in their childhood and youth by the conditions prevailing in their colonies that they either remain there their whole lives long or at least return there when their breeding instinct begins to take over. Their genetic program, however, only enables them to build a brain that is capable of learning for a while. And what there is in the world for these growing birds to learn is nothing more than that they should live in a colony in this particular place. If we raise one of these birds at home from the time it hatches until it reaches sexual maturity, it will be almost impossible to reintegrate into the seabird colony of its parents. That is because it does not have any inborn programming for this and thus will try instinctively to follow us in our world, which it now regards as its world.

With herd animals, like horses or bison, it is basically no different. Who they will run after later depends on who they grow up with at the beginning. A horse that is suckled and raised by a zebra will later on always prefer to join a herd of zebras than a herd of horses. It has no genetic program that tells it: "You're a horse."

Instead the neuronal circuits in its brain are programmed after it is born by the experiences it has during its early development. Its genetic base only determines that it can develop a brain that at the time of its birth does not have completely finished neuronal circuits. How the still-open neuronal pathways that will control its later behavior as a herd animal actually connect up with each other depends on the experiences it has after it is born.

The species-specific song of a songbird, for example, the nightingale, is not an inborn thing. In the nightingale brain there is a region that is responsible for generating its song, and this develops only after it is hatched. The nerve cells of this region first create a great number of extensions and contacts, of which in the course of further development only those are maintained that are stabilized by hearing the species-specific song that the father normally sings repeatedly to the nestlings in the neighborhood of the nest. If during this period the young nightingale repeatedly hears only the crowing of a rooster in a nearby barnyard, then its later song will sound more like a raucous cock's crow than a nightingale song. Nightingale parents instinctively avoid locating their nests where there are likely to be disturbing and alien noises, and the fathers sing most avidly at night, when all the other songbirds are asleep. The complicated song of the nightingale differs from region to region, but in the way described, the young birds always, from the start, automatically learn the home "dialect" sung by their fathers.

Much more exciting than these relatively passively adopted experiences are the effects on developing neuronal circuits in the brain of earlier experiences acquired in the course of an individual's own activity.

The most enduring experiences that a bird or a mammal can undergo are experiences that help it to cope with fear. Every newborn is afraid when it is taken away from its mother. We are all familiar with the cries of ducklings, kittens, or puppies—all of them birds or mammals—at such a time. This fear is coupled with a stress reaction. The transmitters and hormones secreted in the course of this reaction contribute to a process whereby all the neuronal pathways and circuits that the newborn uses to cope with its fear become broken in, that is, reinforced and improved in their efficiency. If the young offspring finds its way back to its mother, the fear is pacified, and all the neuronal circuits in its unfinished brain that were activated in bringing this about have now become better worked out and more effective. Consequently, in the future the offspring will try even harder than before to avoid separation from its mother; it will note the behaviors and means that helped it find its mother again, and it will reinforce and firmly establish all the neuronal pathways that connects it with the protection-providing mother: her smell, her appearance, her behavior. Thus in the future it will recognize its mother a little better and be better able to seek her protection.

The earlier these formative experiences in dealing with fear can be imprinted on the brain and thus the more formable the circuitry of the brain still is at the time these experiences occur, the better they are established for the remainder of the individual's lifetime. They then look like inborn instincts, they can be triggered like inborn instincts, but they are not inborn instincts. Rather they are experiences imprinted on the brain during early childhood in the course of coping with fear and stress.

The more unfinished the brain still is at the time of birth and the slower it develops during the following period, and the longer it takes for all of its neuronal connections to be definitively worked out and established, the greater and richer the opportunities are for the individual's own experiences and the conditions it encounters in its own life to become anchored in the matrix of its brain.

Primates—we humans and our nearest relatives, the anthropoid apes—are distinguished by the fact that we come into the world with a particularly unfinished brain that remains susceptible to formation by experience for a long time, and by the fact that we live in groups that are really extended family units, big families. Every newborn that grows up in such a group is imprinted by factors it finds there that provide a sense of safety and security, just as ducklings are by whatever mother they encounter who offers them protection—without a genetic program having built any particular neuronal connections for this into the brain. But because with the primates this imprinting is significantly more complicated than with ducks, it is no longer called imprinting but bonding.

2.3   Lifelong Programmable Structures: The Brains of People

If you come into the world with a brain whose final wiring—the neuronal circuitry that will determine your later behavior—is only going to be connected up, consolidated, and smoothly established in accordance with how you use it in the course of your subsequent development, you have a big advantage. In order to preserve your inner balance and the inner order you need for your survival, you can no longer rely exclusively on the genetically anchored programs that have arisen over the previous millions of years. Everything you need most critically to survive in the world you are born into—things that are of crucial importance in the place and time in which you live—can be anchored in your brain in the form of specific neuronal connections by the concrete use you make of your brain after you are born. You can benefit from the experiences you yourself acquire with such a brain not only in dealing with your own life later on, but also in creating the conditions in which you will raise your offspring. In this way it actually becomes possible to pass on qualities you yourself have acquired to the next generation. This is indeed an incredibly great advantage, a completely new possibility: passing on the abilities and achievements gained by one generation to the next. This is the beginning of what is known as *cultural evolution*.

An interesting thing about this is that it does not require a human brain. As can be easily shown by experiments, rats can do this too. If you keep rats in a laboratory, you can always find some mother rats who take care of their young in a very painstaking manner, and others who are rather sloppy in this regard, who hardly build a proper nest, repeatedly leave their young alone, and in some cases even devour them. If you redistribute the female offspring immediately after birth in such a way that half the young raised by a "good" mother are her own and half are those of a "bad" mother, all of them will later turn out to be mothers who take scrupulously good care of their young. Conversely, all the female offspring raised by a negligent mother, even if they are biologically from a "good" mother, will grow up to be "bad" mothers. We would normally presume that free-living mother rats who care for their offspring inadequately would not end up having a chance to pass on this particular quality to their offspring over several generations. But this is not necessarily the case. Rats raised by bad mothers, even if some of their siblings are devoured as babies, turn out as adult animals to be more simply structured and more strongly instinct driven than those raised by good mothers. They are more belligerent and brutal and for that reason, primarily in the case of male animals, more successful sexually. The circuitry in their brain is more "primitive," less complex, and not so densely networked. When the need is for fast, unequivocal, and uncompromising reactions, a rat with such a simply constructed brain has the advantage. Since this is frequently enough the case in a free-living rat colony, rats whose main ability is to develop a somewhat more complex, more densely networked brain ultimately really cannot accomplish much. Such rats remain the prisoners of the circumstances they live in, which they are not capable of changing. Even under laboratory conditions, the more primitive conditions cannot be kept from setting in over the long run. As soon as the lab colony gets big enough, the rats with the more simply structured brains again begin to dominate and reproduce more than the others. To give the more circumspect rats a long-term chance, the rules that govern the life of such a colony have to change in such a way that these particularly prudent, learning-capable, and sensitive rats can, like the others, find enough food, ward off dangers, acquire a mate, and raise offspring.

Such changes in the life conditions of rats have never occurred in their entire developmental history. As with the ancestors of other mammals, their ancestors succeeded quite early on in occupying and defending a habitat in which refined brain capacities were not of primary importance and in which they could succeed pretty much just by remaining the way they already were.

Our own ancestors were less successful in this respect. They were not able to conquer a niche in which they could live with any degree of comfort. Their original habitat, the African rainforest, began to shrink rapidly, until only the best and toughest climbers among the primates could make a go of things there. The habitat available outside the forest, the savannah, had already long since been occupied by species that were considerably better adapted to the conditions there. These competitors had already taken over all the possible food sources, they were faster or stronger, and were better at defending themselves or attacking others. Newcomers hardly had a chance in this specialist-ruled world. In order to survive in it, newcomers had to build up and further develop an ability that none of the others possessed. They had to stay together and try to assert themselves as a group, a clan. Only in this way could they take advantage of the different abilities and talents of different individuals and thus achieve as a community what no single one of them could achieve on his own. That was their only chance. But this could only be done by groups whose members felt closely connected with each other and in which each knew the others' special abilities and weaknesses as precisely as possible. Under these conditions, unlike with rats or other species living in groups, it was an advantage to be highly capable of learning, highly circumspect, and highly sensitive—that is, to possess a brain whose definitive patterns of neuronal interconnection remained shapeable for the longest possible time by these humans' own experiences.

In the course of our ancestors' further development, this capacity was quite pointedly favored by selection. However, the evolutionary selection process involved here was not the one that has been known since Darwin's time as the "survival of the fittest," but rather more particularly a second mechanism that Darwin also recognized but that has hitherto received insufficient attention. This second evolutionary selective mechanism is known as sexual selection. Among all socially organized animals with a relatively long developmental phase, the very definite choice of sexual partners who appear attractive on account of specific characteristics and are capable of assuring their own survival and that of their offspring is of great importance in the successful propagation of the species and thus also in the further transmission of the combination of genes that is responsible for those attractive characteristics. In the course of evolution, this selective process, known as partner or mate choice, became more and more important. Along with the selection of specific physical characteristics, it led to the selection of psychological characteristics that proved highly successful in raising offspring and at the same time to selection of the genetic configurations these psychological characteristics were based on. Now the greatest success in the propagation of their species no longer automatically went to those who produced the greatest possible number of offspring. It now went to those with offspring who were the most talented at learning and at bonding, were the most prudent and circumspect, and were the most competent in forming and consolidating social relations within the extended families of these early humans. The more effective the parents were, especially the mothers, at creating conditions conducive for the development in their offspring of these social abilities, the greater the survival chances were for the whole clan.

Criteria for choosing an appropriate mate, in human beings much more than in other animals, were (and still are) determined by the experiences accumulated by individuals, especially in the early stages of their development. Choosing a mate who seems well suited to further these experiences has for an inevitable result that the corresponding genetic configurations of both parents will become stabilized at first in the gene pools of certain families and then, through sexual intermingling, will become established in the gene pools of extended family groups, clans, tribes, and finally the entire peoples that arise out of them.

Advancing socialization brings with it the formation of stable family groups. This is an essential prerequisite for the protection of offspring against all the influences from the external world that might disrupt the maturation of their brains. It also makes possible a high degree of social determination of the developmental conditions within particular family and clan groups. Close emotional bonding between the two parents is the prerequisite for the development of the family and thus for bonding between the parents and their children. As this kind of bonding took place, hand in hand with it there occurred a breathtaking increase in the mental, emotional, and social capacities of the clans that were able to develop this kind of bonding to the greatest extent.

Just how the neuronal circuits that were not yet fixed at the time of birth later actually hooked up with each other and with already hardwired neuronal networks in the brain depended on the concrete experiences that newborns had in dealing with challenges and threats in the real world they lived in. It became essential to hold open and unfinished an ever-increasing proportion of the neuronal circuitry of the brain; but this could only happen if the parental generation could provide their offspring with a sufficient amount of protection from external threats during the period of brain maturation. And this was possible only where sufficiently close bonding had developed between the members of the family, the extended family, and the whole nomadic group. If the bonds among the adult members of a clan were strong enough to ward off the dangers to which the children, with their still immature brains, were vulnerable, then the genetic configurations that produced ever more learning-capable brains were able to persist down through the generations. If the self-assertive, egotistical interests of the adults were too strong for them to provide the needed protection to their young, only those offspring could survive whose brain development was more genetically controlled and whose behavior was more strongly directed by inborn instincts.

At this point during the early phase of human development, a fundamental split occurred. The clans that were unable to develop the kind of emotional bonding we have been talking about could not provide the basic conditions needed for the formation of these brains, which on account of the fact that they were maturing ever more slowly also were becoming ever more adept at learning. Offspring without these kinds of brains were incapable of learning to bond closely with large numbers of their group. Creatures with limited learning ability who were still largely controlled by instinct were unable to make the transition to humanity. There were also those who did make the leap, but when the bonding that had held them together up to that point was destroyed—in most cases by external disruptive factors—either they died out or were able to continue to survive only through the brain development of their offspring being accelerated once again and their behavior again being directed by less complex, more instinct-controlled reactions.

Our own ancestors must have succeeded in maintaining and strengthening the bond between parents and children. They must also have figured out how to make a second and much more important bond stronger and more enduring—they must have succeeded in imprinting on the brains of their offspring the sense of a close bond among the members of the family, the extended family, the tribe, and the entire community, which was continually growing in size. The better able they were to develop a sense of togetherness, the better they became at taking advantage of the mental and physical capabilities and skills of their individual members to strengthen their communities, to uncover new resources, and to fend off enemies. Thus the basic attitudes and shared convictions as well as the goals and motives for action of these early extended families and clans were transmitted from generation to generation along with the knowledge of their circumstances they had acquired and the abilities and skills they had attained. Identification of new generations with the goals, desires, and outlooks of these early human communities was reinforced by handing down traditions concerning the history and development of their ancestors. In ways such as these, human groups became progressively better at exploiting and defending the resources found in their area of settlement, at building stable social structures, at developing historical narratives and traditions that reached further and further into the past, and thus altogether in strengthening the inner bond that held them together and was the basis and motivating factor for all their common achievements.

On the long road of development through the stages of transition from apes to humans, continual small changes in the gene pool took place and were favored by the processes of natural selection, but even more so by the process of intentional partner choice. Among the genetic changes are to be counted the gradual reduction of hair covering, the constant slowing of the speed of brain development, as well as a number of anatomical changes primarily connected with the formation of the pelvis and the development of the extremities and the larynx. These changes made possible the birth of children with larger brains who could walk upright, whose hands were free for manipulative use, and who were able to develop speech. Reduction in the amount of bodily hair strengthened erotic bonding between partners. Naked skin and the frontal sexual union that became possible with the upright positioning of the pelvis were crucial conditions for more emotional and also tenderer sensual encounters between men and women. The fact that partners could now look into each other's eyes while mating and recognize each other personally gave additional strength to the bond between potential parents. Prolongation of women's readiness for mating from originally limited time periods to the entire year in addition to the new emphasis on attractive secondary gender characteristics facilitated the arising of not only intensified but also more enduring sexual and erotic bonds between men and women.

These conditions were not only important for sheer survival. They were also important for the development of densely networked brains capable of lifelong learning. They were also key for the strengthening of the social relations in these early communities that were the indispensable basis for that development. The genetic modifications needed for the anatomical changes that took place—according to the data of molecular biologists concerning the genetic differences between present-day humans and our closest relatives in the animal kingdom, dwarf chimpanzees—involved at most two percent of the entire genetic base. And this process of genetic change was completed around one hundred thousand years ago. The aspect of the genetic base responsible for brain development in humans also has not changed since that time. What have, however, changed a great deal over those thousands of years are the factors that determine how and for what purposes human beings use their brains. These factors are notably: social relations; the knowledge that the acquisition of speech, writing, and data-recording capabilities made it possible to accumulate and transmit; the great growth in communication, which opened up new avenues for transmitting knowledge, abilities, and skills from culture to culture as well as from generation to generation.

Thus the world of human experience determined by cultural development and tradition became more and more complex, multifaceted, and richer. In this world, human beings had the opportunity in the course of their individual development to deal with a great number and variety of challenges. During their lives, they could acquire an ever-greater number of new experiences and anchor these in their brains as patterns of neuronal connections. In this way the thoughts, feelings, and behavior controlled by these neuronal circuits became susceptible of undergoing change into advanced old age.

2.4   Structures for Open-Ended Programs: The Human Brain

It is now already eight hundred years since the Staufer emperor, Friedrich II, showed experimentally what happens to the human brain if its development is left entirely to genetic programming. In order to find out which primordial language the brain would produce entirely on its own, he had two children raised by nurses who were forbidden to speak even a single word to them. For the emperor, the result of this inhumane experiment was quite unexpected. The children did not begin speaking Aramaic or even Greek or Latin. Instead, they became retarded in their overall development and ended up dying. How their brains had developed under these circumstances was not further investigated at the time. Their brains must have been a pitiful version of what they could have become.

Nowadays still, most people on our earth grow up in conditions that lead to their not being able to take advantage of their latent potential to develop a highly complex, densely networked brain capable of lifelong learning. Nowadays still, most people on our earth are forced to use their brains throughout their lives in a very one-sided way for very particular purposes.

This applies not only to those who are occupied, day in and day out, with satisfying their most essential basic needs—getting enough food; defending against life-imperiling encroachments, threats, and illnesses; finding a peaceful place to sleep; and maybe even finding a sexual partner with whom to start a family. It also applies to all those who at one time or another in their lives have found a specific strategy for coping with their fears and maintaining their inner order, and have ever after compulsively used this same strategy in the same way, because they believed it would solve all their other problems too. The neuronal circuitry activated in their brains by this strategy thus becomes progressively better connected and more smoothly facilitated. Initially small neuronal pathways gradually turn into something like solid roads and finally even major freeways. The original coping strategy becomes a well-established program that determines the subsequent thinking, feeling, and behavior of the people in question. They compulsively try to create and maintain the conditions that will allow them to keep proving the effectiveness of this particular set of skills they have acquired. As long as they manage to do this, they get better and better, more efficient and more successful, at coping with certain definite tasks. However, they fail quite pathetically as soon as circumstances change and they are confronted with new challenges that cannot be dealt with by using the same old, well-broken-in neuronal circuit patterns in their brains. Such a lopsidedly programmed brain, used again and again in the same way for the same purposes, also remains a pitiful version of what it could have become.

For example, there are computer addicts who from childhood on have been so intensely involved with their keyboards and their own computer worlds that later on as adults they are hardly capable of carrying on direct conversations or are unable to attract the opposite sex through any other means than the computer. There are mathematical geniuses who cannot tell a seagull from a goose, football virtuosi who can neither swim nor ride a bicycle, and chess masters who can neither sing nor dance.

As these examples indicate, it is not always an advantage to have a brain whose final wiring is determined by how a person uses it, or is forced to use it. What becomes of such a flexible, learning-capable brain and whether or not its inherent potential to form complex neuronal circuitry can be utilized depend on the conditions into which people are born and in which they have to lead their lives. In places where there is not enough to eat, where one's life and family are in constant danger, exchanges with other human beings are reduced to whatever might help to overcome these problems. Where jealousy and mistrust rule and everybody is everybody else's enemy, it is impossible to develop a sense of solidarity. Under such circumstances, exchanges with other human beings are determined purely by the need to assert and promote oneself.

No one is able to choose the circumstances in which he grows up. And no one can control the early experiences that determine how and for what purposes his brain is used—and thus what neuronal circuit patterns get built up and stabilized in it. A hundred thousand years ago, sophisticated speech such as we have now did not exist. The people of those early times had no words for many things about which we now effortlessly communicate, not only in our mother tongues but also in other languages we learn later in life. Their possibilities for expressing themselves about their own individual or general cultural experiences and transmitting knowledge about them were still very limited. There was also no written language that could be used for passing down experience and knowledge from one generation to the next. If, however, one of these early ancestors of ours were to come into the world now, he would be able to speak fluent English the way we do. He could read the things we ourselves read these days. He could even perhaps exchange ideas in German or some other foreign language with people of other cultures—all just as well as we ourselves do today. As we have said, the physiological basis for all this was already present a hundred thousand years ago; only in those days the conditions were lacking for making use of this physiological basis the way we do today.

What drastically changed in the last stage of evolution was not the genetic base needed for developing a highly complex, densely networked brain capable of lifelong learning. What changed were the fundamental conditions necessary for the realization of this potential. These had to be gradually created and maintained over our developmental history from generation to generation. Every step in our development, every discovery, and every invention made by the people of a particular culture helped to expand and extend the way they used their brains. The more use they made of the neuronal connectivity existing in their brains, the more complexity it was capable of developing.

Even today the process of optimizing our conditions for development has not come to an end. In different cultures it has moved with different speeds and advanced to different levels. Historically, the possibilities the people of a given culture had for creating the conditions for their life and development were very strongly determined by the realities they faced in their natural habitat. An increasingly important role came to be played by individual and collective strategies for dealing with life adopted from previous generations—by the knowledge that had been accumulated, the skills and abilities that had evolved, as well as by outlooks and basic convictions that were handed down. Division of labor and specialization led to communities being increasingly subdivided into smaller groups and becoming more hierarchically organized. This is the point at which the quality of brain use began to vary in accordance with social stratum.

At all times during this developmental process, individual members of the community as well as particular strata or groups in society—even entire cultures—ran the risk of overdeveloping and rigidifying already evolved coping strategies that were regarded as especially successful, along with the abilities and skills and basic convictions and values connected with them. When this occurred, it led to a canalization of developmental conditions of the offspring that became more and more pronounced from generation to generation. The limitation of the potential for the offspring's brain use that this entailed favored building up and smoothly establishing particularly intensively used neuronal pathways at the expense of other less frequently activated nerve-cell connections. The better this worked, the more thoroughly the next generation became programmed for the achievement of certain specific goals (of the family, of the clan, of the social stratum, of the society, of the whole culture). Extreme examples of this are still to be found today among many native peoples, for instance, the various tribal groups of

Papua New Guinea who, in fairly complete isolation from each other, passed through a number of very original, in some cases even bizarre-seeming, culture-specific canalization processes. Family-specific canalization processes can lead to highly specialized attainments and abilities in particular occupations (dynasties of artisans, merchants, and public officials), particular arts (artistic families, musical families), or in very questionable fields of endeavor (the Mafia).

However, the advantage gained over a number of generations of increasing specialization always transforms into a fatal disadvantage when external conditions begin to change and other abilities, skills, and ideas become essential. Unavoidably, conditions have always changed and will always continue to do so. On the individual level, they change just because a person gets older, has exchanges with other people, has new experiences, but also because she loses certain skills and has to look for alternative ways of dealing with things. On the level of the family, conditions change through influences on the children coming from outside, and in the case of a marriage, from the influence of the partner's family. On the level of particular social strata and social groupings, change happens through the development of new technologies, the use of new resources, and the changes in the structure of society that these bring. And finally on the level of whole cultures, it comes from increased mixing, contact, commerce, and general exchange with other cultures.

Throughout our history, individual families, social strata and groups, and whole cultures have tried again and again to stave off this process of opening and mixing. But in the long run, it has never been possible anywhere on earth to hold back or reverse this process. Everywhere people, with their learning-vulnerable brains, tend to broaden their store of knowledge, to acquire new abilities and skills, and to have new experiences. And everywhere this knowledge, these abilities, and these ideas are transmitted to other people, adopted by other people, and exchanged with other people. In the past this mostly happened involuntarily and unconsciously (through commerce, wars, migrations, etc.). Today this process of transfer and exchange of information between people of different families, social strata, groups, countries, and cultures can be consciously and purposively arranged. Thus for the first time we can intentionally expand the conditions of life and development that until now have always had a canalizing effect on the final formation of the brain, and in this way we can prevent the facilitation of one-sided neuronal connective patterns in our brains. Only after these one-sided programs that we ourselves have created have been sprung open and undone step-by-step can we take advantage of our genetic potential to develop a complex, densely networked brain capable of lifelong learning. Through this process, we can attain greater subtlety in perceiving and processing changes in our world, an ever greater and more meaningful level of exchange with other human beings, greater efficiency in the maintenance of our inner world, and last but not least, ever better developmental conditions for our children. In the course of human history, prototypes of the ideal kind of brain resulting from this opening process have already frequently appeared on the level of individuals (perhaps when you have finished reading this book, a few examples will occur to you). But this model has yet to be produced successfully on a mass scale.

3

ADVICE
ABOUT
INSTALLATIONS
ALREADY IN PLACE

If you had the brain of a stickleback fish, then you would not have to think about what made you perform the same eternal mating ritual with your partner every spring. It would not even bother you that if you ever happened to be disturbed in the middle of this ritual, you would have to start the whole thing over again from the beginning in order to bring it to a successful conclusion. And then when your children had grown up and were entering their mating phase and had begun to perform the same dance as you had, you would not wonder how they came to master this complicated ritual without ever having seen or practiced it. You could simply rely on the fact that all the nerve-cell connections that a stickleback needs in his stickleback brain to survive and reproduce are already installed there in the same way they always have been. The genetic programs that bring this about are as old and unchanging as the stickleback itself and its remarkable mating dance.

If you had the brain of a grey goose and had been raised from the moment you hatched by Konrad Lorenz, then every spring you would perform the same genetically programmed mating ritual as all other geese perform, except you would not do it in the presence of a goose or a gander, but in the presence of Konrad Lorenz. Despite the patent futility of your efforts, you would still try the next spring in the same way again to move him to mate with you. At a certain moment in the past, when some of the neuronal circuits installed in your brain had not yet fully matured, the image of the old man was firmly anchored in your duckling brain, and this pattern of neuronal connections was put as firmly in place as if it had been installed there by a genetic program.

With the brain of an ape, you would acquire a few more experiences than a duck and anchor them correspondingly in your brain. Later on you could maybe even undo a few of your experience-determined neuronal circuits or overlay them with new installations. But you would not have the understanding to assess whether or not the installations already in place would continue to be useful to you or not. You would have no idea whether you should hold onto them or not. In order to choose whether neuronal connections already set up in your brain to control your feeling, thinking, and behavior ought to remain there for the rest of your life or should be changed, you need a brain that is capable of continuing to learn for the rest of your life. In case you are of the opinion that you do not have such a brain, you can stop reading this user's manual right now.

But if you have decided to read further, you should still pause at this point for a moment and think about what it might mean to you if you were to find out that the neuronal circuitry installed in your brain already is not really running as ideally as might be desired. This will involve feelings of guilt and blame and the question of how to deal with the realization that, without any intentional input on your part, before you were born, during your childhood, and in the period of your life that followed, certain neuronal connections were set up in your brain that continue to play a decisive role in determining how you think, feel, and behave right now.

Who or what do you want to blame this on? Your genes, the chance combination that your parents started you off with? The family situation you grew up in, as a result of which you were able to have (and have anchored in your brain) only certain specific—perhaps quite one-sided—experiences? Or the social circumstances you were born into and grew up with—the culture, the time, the region—that have largely determined, and perhaps even seriously limited, the developmental conditions of your life right up to the present moment and thus have also limited your potential for using your brain?

In totally different times, in a totally different culture, with totally different parents, you also would have been quite different. You would have had a different brain and thus would have thought, felt, and behaved quite differently from the way you do now. You would have been so different from the way you are today that the chances you would even recognize yourself are slim. All the elements that compose your personality, the things that you are proud of as well as the things you do not like about yourself—that you even suffer from—your weaknesses and your strengths, your abilities, your knowledge, your desires and expectations, even your dreams and fears, are the result and expression of the patterns of neuronal connectivity that have been put in place so far in your brain. You are a product of chance occurrences, of factors blended in your genome by chance, and of chance conditions in which you were able to make the most of some of these genetic factors and not able to do well at developing others.

As Baltasar Gracián said, "Water takes on the good and bad qualities of the strata through which it flows, as human beings do of the climates in which they are born." Trying to assign blame after the fact, trying to decide who or what is at fault for the mental, social, or familial climate in which we were born, makes little sense. Such a retrospective effort can only serve one purpose, and that is to discern how we ourselves have participated in producing a particular climate so that we will not be guilty of the same in the future out of ignorance. Because the climate in which we continue to grow and live from here on out is just as subject to change as the way we use our brains.

3.1 Optimally Successful Installations

In order to make the most of its genetic potential to develop highly complex and lifelong-alterable neuronal circuitry, a human brain requires optimal conditions. This is true even before birth. During the intrauterine period, when the brain is growing very fast, no changes must occur that disrupt the availability of the basic materials needed for this rapid growth, and the balance of substrates, cofactors, and other substances that influence the brain's maturation process must be right. Of concern here are not only inadequacies in placental supply or metabolic disturbances in the mother, but also ingestion of substances by the mother (alcohol, nicotine, drugs, medicines, etc.) that can reach the developing brain of the fetus through the placenta and change the complex of conditions that control the maturation process that is going on. Changes in the concentrations of certain hormones, growth factors, and other signal substances circulating in the mother's blood, which can be triggered by psychological or physical difficulties during pregnancy, can also influence the development of the fetal brain.

Toward the end of pregnancy, the various sense organs and the neuronal pathways associated with them in the brain have already developed to the point where the fetus can have its first sense perceptions. It feels rocking, its tastes the amniotic fluid, and it hears the mother's heartbeat and other sounds, such as voices and music coming from outside. Anything that penetrates into its world that it can perceive is connected by the unborn child with the sense of security that normally prevails in this inner place. Sudden, possibly frequently repeated, disturbances during the pregnancy, such as loud noises or moments of fear and stress on the part of the mother, which the fetus perceives as changes in her heartbeat and which are accompanied by changes in the maternal blood supply and by secretion of a variety of hormones, can lead in many children to their sense of security already being quite reduced by the time of birth. Such children come into the world already anxious and fearful, and are much harder to calm and pacify through motherly

attention than babies who were spared these intrauterine disturbances.

Every human being experiences his or her first deep fear and
stress reaction at the time of birth. In the desperate state following
this dramatic change in its world, the child must try to find a way to
recover its lost inner balance. The most important experience that a
newborn can and must have during the first days and weeks in its
new world is one that becomes anchored in its brain and has a decisive effect on its subsequent development. This experience is the
feeling that the child is capable of coping with its fear. In order for
it to acquire this feeling, the newborn must be able to express its
fear. Then it depends on its crying being heard, on someone (normally the mother) giving it attention, rocking it, taking it to the
breast, talking to it, warming and soothing it. Only if the baby finds
somebody who enables it to sense and perceive as much as possible
of what it is already familiar with from its time in the mother's body
and who it associates with the sense of security it experienced there,
can it overcome its fear and recover its inner emotional balance.

The more frequently it is successful at this, the more deeply
this experience that it can cope with its fear with its mother's help
becomes anchored in its brain. Its self-confidence grows in this
process, along with its confidence in its mother's ability to provide
a sense of security. The child develops a close emotional bond with
the mother (or with another primary caregiver), and in the course
of its subsequent development, it takes over from her all the abilities, skills, ideas, and attitudes that seem important to it for dealing
with its own life. It also broadens the emotional bond it has with
its mother and extends it to the persons who are important to her,

to those with whom she is emotionally close and in whose presence the child also feels safe and secure. This usually happens first with the father. Later come the grandparents, relatives, and other people who are close to the parents. The child also appropriates their abilities, attitudes, and ideas. The closer the child feels to these other people, the easier this is and the better it works.

During this phase, things are not that different for the child than for a sprouting seed, which sends its ever more densely branching roots down into the earth and anchors them more and more solidly there in order to gather the nutrients it needs for the development of its stem and leaves. Children can only grow such roots if they are given the opportunity during the first year of their lives to develop close, sure, and firm bonds with other people, as many people as possible, people with different abilities, ideas, and talents. In the case of a seed, genetics decide whether it will grow a deep-reaching taproot or a more horizontal, widely branching fibrous type of root. In the case of children, very deep but only very slightly branched roots always form when the environment they grow up in is defined by one person or very few of the same kind of people. They form more widely branching roots that are closer to the surface when they have relationships with a large number of different people who provide them with only a little sense of security.

In order not to be blown over by even a small storm, trees on swampy ground need the deepest possible roots, and trees on stony ground need more horizontal, spreading roots. What children need are roots that will hold fast anywhere, in any kind of weather. From Africa, the cradle of mankind, comes a very ancient piece of wisdom, which in one sentence summarizes the conditions children need in order to take full advantage of their genetic potential for developing a complexly wired brain that is capable of lifelong learning. An African proverb says: "To raise a child properly, you need an entire village." In a village community, children find many quite different stimuli and challenges. This is what they need to acquire a broad range of varied skills and to set up in their brains the neuronal connections that these skills activate. In a village, children can enter into a growing circle of strong, safe bonds with very different kinds of people and experience the feeling that within this community, protection and security are available to them.

Villages that provide this sort of experience have grown increasingly rare, even in Africa. And if villages like this still exist somewhere, what they offer children is not enough for them to develop what they need these days as urgently as roots: wings. Today children need wings to fly beyond the boundaries and limitations of the community in which they happened to grow up. Such wings as this do not just grow by themselves. Children who do not feel secure in the world they grew up in are afraid of flying. Their taproots are so strongly attached to the few bonds they have that they cannot get off the ground. And the ones with the wide-spreading fibrous roots are all too apt to take off and soar away before their wings are developed enough to control the direction of their flight.

Generally, it is impossible to tell whether or not a child is capable of growing sufficiently strong and guidable wings until it has grown up and begins to use them. In the case of rats, on the other hand, which have far less flexible and learning-capable brains than humans, we can see clearly what happens in their brains when they are raised in conditions that make it possible for them to grow the very tiny psychological wings rats are capable of growing—we might call them rat wings. Under the right conditions, their cortex becomes thicker, it contains more synaptic contacts, its nerve cells have longer extensions with more branches, there are more glial cells, and the cortex even has more blood vessels with more branches to supply blood to these more complex neuronal circuits. As adults, "winged" rats can handle more difficult tasks in a more skillful fashion. They are more competent and have less fear of new things than their cousins who have grown up under "normal" conditions, in the usual kind of cage, and did not have the chance to experience extended family groups, to enter into many-sided contacts with other group members, to dig burrows, and in general to discover a more colorful rodent world with all kinds of different challenges and stimuli. But the most important thing of all does not become visible until these animals have grown old, that is, until after two years have gone by. Then we find in the brains of the "normally" raised rats a considerable number of degenerative changes; whereas the brains of the "winged rats" still look quite normal. Some brain researchers call this "the Matthew principle," based on the phrase from the gospel, "For whosoever hath, to him shall be given" (Matthew 13:21).

3.2   Defective Installations

At the time of birth the human brain is still quite immature. The only neuronal connections that are adequately developed are those that are absolutely essential for survival during the first phase of life—those required for regulation of basic body functions, for processing essential sense perceptions, and for coordinating the first motor reactions. Their most important overall function is to set in motion a reaction suitable for restoring lost inner balance when threats and disturbances occur. The more clearly a child is able to express either its discontent over an unsatisfied need or its contentment over a satisfied one, the better this process works. The first kind of expression normally brings someone to its side who will help. The latter kind reinforces the likelihood that willingness to provide help will continue to be there in the future.

These two skills are not equally present in all children. And not every mother is capable of accurately interpreting the signals through which her child indicates the state it is currently in. All mothers are also not equally good at recognizing and eliminating the cause of their baby's current problem. All mothers are not equally good at recognizing the joy a baby expresses when it has succeeded in recovering its inner balance, and what is more important, all are not equally good at providing a reaction that their child can recognize and that responds to and reinforces its joy.

There are children who enter this world with much more fear than others. And there are children, who following their birth, encounter conditions that are not conducive to developing a sense of security. These children have less confidence than others about their ability to eliminate a disturbance to their inner balance through their own effort and with the help of their mother—and less confidence that they can share their joy over this successful enterprise with her. There are psychologically disturbed mothers, immature mothers, unhappy and discontented mothers, insecure and fearful mothers who are plagued by self-doubt, moody and fickle mothers, overly self-centered mothers, or mothers who are overly controlled by others. There are troubled and overburdened mothers, hard and insensitive mothers; there are disoriented mothers and mothers seeking orientation—there are simply all kinds of mothers who cannot provide the conditions their children need for optimal development. Instead of a secure, supportive bond, there arises between them and their children a very uncertain kind of bond. The child is either too much clutched and clung to and thus hindered in developing its abilities, or it is left too much to itself and receives too little stimulation and guidance in developing its abilities.

The consequences for the further development of a child's brain of this kind of uncertain bond with the primary caregiver can become serious and long lasting if the child does not subsequently have an opportunity to develop close emotional ties with other people. This is usually the case if the mother herself has few such bonds with other people, for example, if her relationship to the child's father and to her own father and mother and relatives is also uncertain and insecure and she has failed to develop close and firm emotional bonds with other people beyond her family. If the mother herself, as the primary caregiver, is not embedded in a system of security-bolstering relationships involving a large number of different kinds of people, the risk is high that a very one-sided basic pattern of thinking, feeling, and behaving will emerge in the child under the exclusive influence of its unfortunate mother, and that the neuronal connections that go along with this one-sided pattern will become established and consolidated in its ripening brain.

But if the child succeeds in finding people beside the mother who are helpful to it in overcoming its fears and achieving a sense of security, then the basic attitudes, abilities, skills, and the emotional bonds of these people will be adopted by the child and become anchored in its brain. This is the only way of avoiding excessively one-sided and canalized developments that come entirely from the primary caregiver as well as the corresponding early programming resulting from the neuronal connections this kind of development establishes in the infantile brain. In this process, it is important that the canalizing influence of a secondary caregiver be clearly differentiated from the canalizing influence of the primary caregiver, that is, the mother. The best person for this secondary role is the father. But just as not all women can be ideal mothers, not all men are equally able to be loving, empathic "program-opener-uppers" in their children's brain development, a function that is accomplished by providing their children with an opportunity to discover a world that is different from their mother's world. Often what fathers provide is a contrasting program, either presented as an alternative or imposed in an authoritarian fashion, which instead of offering a synthesis, forces the child to make a drastic decision: to use either its feelings or its rational mind, to orient itself inward or outward, to remain in a state of dependency or step beyond all bonds and become autonomous.

How the child decides in the question of whether to adopt the behavior pattern put forward by the father or the mother depends on which one of the two it feels safer and more secure with and which one's strategies seem best suited for dealing with its life, coping with its fears and insecurities, and maintaining its inner balance. Frequently all the requirements are not met by one of the parents alone. Then the child has to make its way through a jungle of conflicting feelings. In some cases, for example, there can be affection for the warm-hearted mother, who (according to the father) cannot do anything, conflicting with a sense of dependence on an overly powerful father, who (seemingly) can do anything. In such cases, premature detachment from both parents usually ends up being the only way out of this jungle.

If one of the parents offers both of these things, or the ideas, basic attitudes, and competences of the two are not essentially distinct, the child runs the risk of adopting their ideas, attitudes, and competences without entertaining any alternatives, even though they might well turn out later to be inadequate, or even highly obstructive, in following its own path in life. A child can escape this danger of prematurely setting up definite patterns of perception and psychological processing only if it can find other people in its life who also provide a sense of security but who think, feel, and act differently than its parents do—who possess different knowledge, have had different experiences, and have developed different skills.

However, most children grow up in family, village, religious, or cultural communities whose members share very particular—often one-sided—ideas, outlooks, and attitudes. Members of these communities frequently have only pretty much the same, limited knowledge available to them and have developed certain skills at the expense of others. Children in such communities can only acquire a sense of security and deal with their fears by adopting the thinking, feeling, and behavior patterns the community members pass along to them. The neuronal connections and circuits activated repeatedly in their brains in this way become more and more rigidly established. The earlier this kind of programming takes place, the more completely it determines children's lives and the harder it becomes for them to undo it later in life.

The danger of building up these very one-sided neuronal connective patterns becomes great if the associated strategies for coping with fear are employed by a person in the early stages of development and subjectively assessed by that person as highly successful. We see certain coping strategies being built up and overused to the point of psychological dependency, yielding such results as, for example, obsessive preoccupation with career, success, prestige, finery, pleasure and distraction, and gambling (excitement). Frequently also, the calming effects of eating are used to cope with fear and developed to the level of dependency (bulimia, anorexia). The same is true of drugs and medicines, which on account of their fear-decreasing, sedative, or euphoric effects are used to calm fear.

The more restricted the range of coping strategies a person acquires early on in life, the greater the likelihood of failure when it comes to dealing with new kinds of psychological conflict and stress. People dependent on a restricted range of strategies are often incapable of finding new ways to meet new challenges. For this reason they tend to deal with their fears and the uncontrollable stress reactions provoked by them by falling back on strategies that in their eyes are proven, but in the eyes of outside observers, often appear incomprehensible. Following this pattern, many people reach the point of retreating more or less obviously into self-created worlds that seem to offer security because they contain certain aspects of early coping strategies that these individuals still see as successful. Other people facing conflict situations tend toward active, outwardly directed solutions, and have recourse to their old, power-appropriation strategies (fits of anger) or to showing off status symbols.

However, the chances are normally quite slim of actually managing an uncontrollable fear and stress reaction using one of these early established coping strategies that are now activated quite unconsciously. Thus many people facing overwhelming and unmanageable problems try to create situations that they actually *can* control with the help of their old coping strategies. Just which concrete situations a particular person will try to conjure up through his own actions in order to prove to himself once more that he is able to deal with the problems resulting from them depends on the experiences he has had previously in coping with fear and stress. Many people manipulate situations into a place in which they can arouse the helping instincts of others. Others' maneuvers aim at reconfirming that they can cope on their own. And still others try through creating certain scenes to convince themselves that even if beloved, security-providing "significant others" reject them or they run into fresh evidence of their own incompetence, they couldn't care less, they can deal—no problem.

These kinds of behaviors, if they are frequently enough employed to cope with fear and continue to be viewed subjectively as successful, become entrenched to the point of dependency and their user compulsively sets them in motion whenever he feels threatened or insecure.

4

REPAIRING
FAILED
INSTALLATIONS

*I*f *you had the brain of a mole, you would not be able to conceive of* what somebody was talking about if they told you about bright sunshine, meadows full of flowers, and colorful butterflies fluttering around just a few inches above your head. You would just turn away and continue tunneling away in the dark the way you had been, with your feeble eyes and the correspondingly underdeveloped visual cortex of your mole's brain. You would not even try to imagine how things look up there, just a short distance above you.

But you do not have the brain of a mole, and even if, on account of a genetic defect or other damaging factor, you had come into the world blind, you could still somehow picture in your mind how the multicolored fields and butterflies around you might look. You would never have seen any of this, but you would have heard about it from other people and later on you would have read about it in your braille books. Instead of burrowing back into your inborn dark world, you would try somehow to make the invisible visible by heightening your senses of hearing and feeling and through your imagination. And the regions of your brain that you used particularly intensively in this process would also develop a more complex and refined form. For in contrast to a mole's brain, a human brain is even capable of compensating for a genetically caused installation failure.

As in all other areas of life, a repair job in the brain works better the sooner it is undertaken. But even if blindness sets in at the adult stage, profound changes are possible in the neuronal circuits responsible for perception and the processing of sense impressions in the adult brain. With the help of certain technologies, it is possible to represent these changes graphically. A particularly impressive example that we can observe by this means involves changes in the area of the cerebral cortex that processes sense impressions coming from the fingertips. This area begins to expand when, after being blinded, a person starts using his fingertips to learn braille. People with a particularly acute and well-preserved sense of touch have the easiest time learning braille. Not only the sensory networks in the cortex, but also neuronal networks responsible for the coordination of movements can adapt to changes in their use. Examples are changes that occur after members, for instance, the index finger, have been amputated, or after strokes that lead to partial paralysis, for instance, paralysis of the right arm. After the loss of an index finger, further development and improvement occurs in all the neuronal connections that guide the movements of the other fingers, especially the thumb and middle finger, which are now used more intensively. And when the right arm can no longer be moved, the areas of the cortex that coordinate the movements of the left arm are modified to the point where the person concerned can write as well left-handed as he could right-handed before. Of course it would be even better to keep the still-usable left arm from taking over all the tasks the right arm can no longer perform. Lately this has been tried in certain rehabilitation programs, and to the great astonishment of brain researchers, it led to hitherto inconceivable new neuronal connectivity in the brains of participating patients. As a result, their paralyzed arms became usable again for many tasks, even if not for all.

This latent, use-dependent plasticity of the human brain manifests not only when failures and losses occur. Especially intensive use of individual regions of the brain can lead to the neuronal networks responsible for associated tasks becoming more complex, denser, and sometimes even bigger. Thus the region responsible for directional orientation in the brains of London taxi drivers (who obviously use this region very heavily) has been shown to become progressively bigger the longer they pursue this occupation.

The brain researchers who found this out were quite amazed to discover how great the use-dependent malleability of the human brain is. If we take their findings to the logical conclusion, the implication is that the brain takes on the form of how it is used. The neuronal connections that we activate especially often and successfully in dealing with the world become more and more strongly developed, and the ones that are employed only quite rarely either stay the way they are or gradually begin to deteriorate. Since there are no two people who have had exactly the same experiences in their lives and have used their brains in exactly the same way, every brain—given the course of its particular history—is unique. And since any person can choose at any time in his life to use his brain a little differently in the future than he has up to now, he may well be able to repair deficiencies that have hitherto developed in his cerebral installations. Most of these deficiencies have become solidly anchored in the brain through repetition of previously adopted strategies of perceiving, thinking, feeling, and behaving that either have been regarded as right or that have never been seriously questioned. There is only one way to get your brain back to the point where you can see as well as feel, smell as well as hear, dance as well as play music, and think rationally as well as sense things intuitively. This way, which has been known to us human beings for thousands of years, is expressed in an ancient Chinese proverb: "Not where you have already attained mastery, should you exert yourself further, but there where such mastery has still yet to appear."

4.1 Imbalances between Feeling and Intellect

There are people who are so dominated by their feelings and emotions that they are hardly accessible to rational arguments. They make their decisions "from the gut." Their relationships with other people are either "heartfelt," or they feel they "take them to heart." It is hard for them—and they often categorically reject this possibility—to solve a problem by means of factual analysis through the use of their intellect. People who work that way are repugnant to them. They are proud of being "feeling people." They are quite happy about being the way they are and have little inclination to think about how and why they became that way.

It does not help much to recall to such a person that perhaps during his childhood he had a very close relationship with a primary caregiver (in most cases his mother), who provided him with a very strong sense of security, not because she knew so much or could do so much, but just because she was always there; and when there were problems that as a child he was already capable of solving, she still hovered over him like a protecting hen. It is equally futile to try to talk to such a person about the undesirable consequences that might come about if all people were as strongly dominated by their feelings as he or she is—to point out that there might be people whose hearts were full of hatred, greed, envy, and jealousy, and who raped, mutilated, or killed other people based on the feelings they had "in their gut" and got tremendous pleasure out of it.

It is so difficult to get anywhere through logical argument with such "feeling people" because they have missed the experience— and thus have not had it anchored in their brains—that problems can be solved through the intellect. They can acquire this experience only if they encounter someone who will help them to have fun again, like a three-year-old child using their intellect to understand what is happening around them and inside them—to become intrigued with this in a fresh way. They must not be dogmatically taught, but rather encouraged, to go out and not just perceive the world but to discover it for themselves. They must be provided with an opportunity to acquire knowledge that helps them see behind the surface of things, to recognize hidden, hitherto invisible situations, and in this way to get better at dealing with the world than they have been up to now. They can only be helped in this by someone who has developed thinking and feeling equally well and who can use both at the same time.

Things are not really much different for people, who on the basis of their personal experiences, usually quite early ones, have come to the conclusion that their intellect, the knowledge they have acquired, and the intellectual skills and capabilities they have developed are the only thing they can rely on. Such people usually categorically reject anything that comes from the heart or "the gut." They distrust their own feelings, have little understanding for the feelings of other people, and try to manage everything "with a cool head." They are usually very proud of their intellectual capacities, regarding them as a special gift, but left to themselves are as little inclined as "feeling people" to spend any thought on why and how they became that way.

But it is easier to talk to these "intellectuals" about this. Often they find out for themselves that during their childhood, there was somebody (usually their father) who made a big impression on them, because he seemed to know everything, could analyze things with great clarity, and was seemingly able to deal with the world in an exemplary fashion using his intellect. What they do not like to think or talk about, however, are the reasons they are afraid of their feelings and emotions and the question of when they began to control with their intellect—that is, actually to suppress—these feelings coming from their "gut" or their heart. Sometimes it is possible to get such a person to acknowledge that there was somebody who hurt his feelings, somebody who right at the beginning was very close to him, with whom he felt quite safe and secure, and that since that time he has come to disdain this person (usually a mother belonging to the "feeling people" group), has rejected her, and perhaps even hates her. How this all came about, he does not know. He only knows that that she got on his nerves tremendously with her sentimentality—this he can clearly recall.

It is quite difficult for these people (mostly they are men who are quite successful professionally) to realize that it was they themselves who broke this close bond with their mother—at the point when they began to notice that she knew and understood too little to be able to make a success of herself out there in the world beyond the family. Because disappointed love is a feeling that cannot be tolerated, these people attempted to suppress this feeling with the help of their intellect.

The better job of suppression such a disappointed person is able to pull off, the harder it is for him to summon up these buried feelings later on. And when they do suddenly come up for some reason (usually in the throes of a sexual relationship), for the most part such men are incapable of dealing with them. What they have to learn again is not that feelings exist, but that it is possible to allow them to be there. They must learn all over again not to be afraid of their own emotions. For that, they need someone who will provide them with the experience that the ability to feel and to express feelings enriches their life, makes it more varied and colorful, and makes them personally richer and more lovable people. Again, help of this sort can only come from people whose ability to feel and think have been equally well developed.

Without support of this nature, both for "feeling people" and "intellectuals," there is only one last-ditch chance of breaking down the one-sided neuronal connective pattern that has developed in their brains—a serious psychological crisis. In the course of such a crisis, destabilization of established connective circuits occurs in the brain as a result of the triggering of a long-term stress reaction. Such a crisis sometimes provides a chance to leave behind old, rut-like patterns of thinking and feeling. But it can also all too easily endanger the person's whole inner order. If a person does not succeed at all, or is not fast enough, in reworking the neuronal connections that have determined his thinking, feeling, and behavior in the past, this process of destabilization can turn into a life-imperiling (or at least illness-producing) threat.

4.2   Imbalances between Dependency and Autonomy

When children come into the world, they are dependent on the help of adults. They need someone to keep them warm, fed, clean, and to look after them generally. And whenever they feel fear, they need someone who stands by them and shows them that it is possible—and later *how* it is possible—to overcome this fear. If a child is lucky enough to find someone who is always there for her when she feels fear and provides a sense of safety and security, the neuronal pathways that are then activated in her brain become firmly established and smoothly facilitated. This is the way a close bond with the primary caregiver arises.

Many mothers are aware of this and strengthen this bond through play—by, for example, hiding for a short time, but then moments later, just when the child is getting afraid, turning up again. When children have the feeling aroused in them that they can bring the disappearing mother back through their own reactions, their confidence in their own ability to control threatening situations grows. At the same time, the neuronal connections activated in this process are established and facilitated. In this way, self-confidence arises, confidence in the child's own competence in coping with problems. In the course of further development, the circle of people who relate closely to the child and provide a sense of security broadens. The child adopts all the capabilities, basic attitudes, and behavior patterns that these people possess that it finds important in maintaining its inner order. Then as the child broadens his own knowledge and capabilities and increases his experience, these early bonds lose their original security-providing significance. This development intensifies dramatically during puberty, when the production of sexual hormones leads to profound changes in the body and in thinking, feeling, and behavior. By the time this process of development is over, the initially entirely dependent baby has turned into a self-determining human being who is part of a complex net of social relationships.

Unfortunately, things do not always work out this way. There are more than a few adults around who during their childhood and adolescence were unable, or did not have the opportunity, to achieve adequate skill at accumulating varied experiences of their own. As a result, they never gained the self-confidence they needed for autonomous development. Such people either stay put in a dependent relationship with their primary caregiver or seek out a partner with whom they can continue to have the same kind of dependent relationship. If they have children, with them too, they develop a dependent and dependency-building relationship based on clinging. What people of this kind need is somebody who will give them the courage to discover their own social competencies and to put them to work in the further shaping of their lives.

Another extreme example of a skewed balance between dependency and autonomy is found in people who during their early development were unable or did not have the chance to build up stable, secure bonds with their primary caregivers. A fundamental cause for this can be abuse, or more often, just neglect.

When a child gets into a situation that starkly conflicts with all its previous experience and in which all the coping strategies it has applied successfully up to that point fail to work, it can lead to psychological trauma. This presents the most extreme case of uncontrollable stress a person can experience. Such a result very frequently occurs when girls get into a situation in which, with the mother's acquiescence, they are abused by their father or another closely related person.

After a traumatic experience of this nature, if the child does not succeed in somehow putting a stop to her uncontrollable stress reaction, then she is lost, for the process of destabilization that has been triggered can take on life-threatening proportions. Every traumatized child feels this, and will thus try with every means at its disposal to get this traumatic experience and the memories of it that keep flaring up afterward, under control. Proven strategies it has used in the past to cope with its fears will now be tried repeatedly, again and again to the point of absurdity, against the trauma it has experienced. In the case of the abused girl, she can no longer rely on parental support. She loses any faith she may once have had in an outside divine force just as surely as she loses faith in her own power. The only strategy that can still provide her with relief is disconnecting the traumatic experience from her memory bank, separating it out by specifically altering her perception and associative processing of the phenomena of the outside world. She is forced to build up defense mechanisms against the memories of the trauma that constantly keep coming to the surface. If she does find a strategy that enables her to get the traumatic memory and the accompanying uncontrollable stress reaction under control, the destabilization process comes to an end, and then all the neuronal connections that were activated by this successful way of coping with the fears triggered by the traumatic memory become reinforced and firmly established. Through repeated successful use of these neuronal connections, there then arise at first small and then increasingly broader and more effective "bypasses," "detours," "prohibited zones," and "rest stops" in the road system of her central nervous system. Many abused children succeed in suppressing the trauma they have experienced in this way. Many split themselves into two personalities, only one of which has been abused, while the other remains unscathed. Many regard the abused areas of their bodies as no longer belonging to them and lose all sensation there. Many degenerate into stereotypic movement patterns or repeatedly try to damage themselves.

Solutions like these are usually found more or less quickly and intuitively, but before the neuronal connections associated with them in the brain have become adequately worked out and facilitated, months and years can go by. But then, eventually, this neuronal facilitation process can work so deeply and pervasively that the memory of the traumatic experience can no longer be retrieved at all.

Such defensive strategies as these are individual solutions that are necessarily clearly distinct from the "normal" coping strategies of nontraumatized children. As a result, traumatized children tend to end up socially in a world apart, and are often labeled by others as psychologically disturbed or antisocial. In this way a vicious circle closes on them, and even after such children have grown up, they are unable to find a way out of that circle on their own.

A second, definitely more common cause of bond disruption in childhood is a lack of emotional attention and love. There are many parents who remain largely concerned with themselves, for whom their professional career is extremely important, who are trying to improve themselves and realize their potential, to broaden their experience, and enjoy their lives as much as possible. They are very concerned with their appearance, their hobbies, the decor of their apartments, and are busy accumulating and showing off a variety of status symbols. Children are often a hindrance to these self-absorbed parents' plans, and with their need for attention, security, and affection, the children all too easily become a burden to them. For the most part these parents do their duty, or at least what they consider to be their duty, and sometimes they even do quite a good job of it. They make sure their children have good food and keep clean and live in hygienic conditions, that they have appropriate, stylish clothing, and all sorts of external paraphernalia and possessions that they think are important for them. They salve their conscience by spoiling their children to the best of their ability. But what their children really need, namely, for the parents to be entirely there for them, to give them full and complete attention emotionally, intellectually, and physically when they feel insecure and afraid, these parents do not provide for them, or at least not when the children truly and urgently need it. As a result, such children are often forced very early on to begin relying solely on themselves.

In such cases, the emotional bond with the primary caregivers is inadequately developed. The children are obliged to compensate for the lack of emotional security that results from this by increased self-absorption. So they create for themselves their own self-determined little world and shut out external influences and stimuli that do not fit in with their fantasies. In this self-created world of theirs, there are no longer any real challenges. It becomes impossible to have the new and varied experiences that should be becoming anchored in their developing brains. Developmental processes important for the child's brain either do not take place or do so only in a limited way. For the child's learning processes, this represents a step backward in terms of motivation, understanding, memory, and recognition of relationships. It also results in a decreased ability to acknowledge and resolve conflicts. The social behavior of such children is marked by their tendency to retreat increasingly into self-created worlds, by rejection of outside ideas, and aggressive defensiveness on behalf of their own outlooks and attitudes.

Most of the time what we find in such cases are very rigid, one-sided, pseudoautonomous strategies for coping with fear. The earlier and more frequently these strategies are applied, the more entrenched the neuronal connections they activate become. These connections can end up entirely determining the feeling, thinking, and behavior of these children. The children then progressively shut themselves off from the ideas of others, particularly adults. Their meager empathic ability prevents them from acquiring a significant repertoire of different kinds of social skills. This results in their not having the kind of foundation that would enable them to work together with different kinds of people in seeking sound solutions for their problems. Thus it is hard for them to reach the point where they can take responsibility for themselves and others.

The effects of early bonding disruptions on the development of the brain and the personality are difficult to correct later on in life. People who were abused or neglected in childhood are afraid of physical and emotional contact. If they do not succeed in overcoming this fear, they remain isolated throughout their lives. They become ego-centered and incapable of bonding. Many have the good luck to find a person who understands them and helps them gradually to enter into relationships with other people again, to regain their confidence in personal bonds, and to allow themselves to work with other people in finding shared solutions to problems. However, many such people are eventually destroyed by their autonomous coping strategies. They become ill and undergo a serious psychological crisis. Under the best of circumstances, this crisis gives them a chance for a new start. But even this new start can work out positively only if there is somebody to help them who himself has found a balance between autonomy and dependence.

4.3   Imbalances between Openness and Self-Differentiation

Many children come into the world possessing an incredible openness; many others develop this quality only after birth. Even as infants they are extremely alert and receptive, extremely interested in anything new. They do not sleep very much and seem to fear nothing. They have a tremendous urge to move around and to express whatever is going on inside them. These children take in more than they can actually process, that is, more than they can fit together into consistent inner pictures of the outer world surrounding them. Thus they are in great danger of drowning in the flood of incoming information they are faced with, especially before they are able to assess its significance. If their parents are not able to create a sufficiently ordered and structured environment for them, their particular gift can turn into a disaster. The strategies these children develop to defend against such a surplus of stimuli range from idiosyncratic behaviors to attention disorders, and even to the development of hyperkinetic syndrome.

The only way to deal with the greater openness of these children is through giving more definite structure to their everyday life situations, not in order to reduce the child's experience, but to provide it with the possibility of ordering and sorting all the input that is pouring into its brain at once. A more structured environment gives the child a chance to get this input under control.

We find the opposite extreme in people who—also often already as newborns or babies—stand out from others by how extremely little they seem to be impressed by what is going on around them.

They seem to be somehow absent and turned in upon themselves. They have very little curiosity and are very difficult to get excited about anything. Neither their need to move around nor their ability to express what is going on inside them is very pronounced. When they play, they prefer always to repeat the same thing; and when they move, they would rather not move too wildly or impetuously.

These children are not particularly fearful. Rather they give the impression of resting unshakably within themselves, like a rock in the surf. This too is a special gift, but being overly closed can have just as detrimental an effect on later development as being overly open. Overly closed children are in danger of taking in too little from the world. And since then also, as a result of this, very little happens in their world, they very seldom come to the experience that it is important for them to gain their own social skills and to get the support of other people in solving their problems. They also rarely have the experience of being important to other people. Thus they develop neither sufficiently strong bonds with other people nor much in the way of social competence of their own. What can help them, and what they need, is exactly the opposite of what is needed by children who are too receptive and open. They need people who provide strong stimuli, who bring a bit more chaos into their world, a bit more unpredictability, irregularity, and the absence of struc-ture. They need people with whom they keep having experiences that get under their skin, that shake their emotional balance and force them to seek out new solutions.

5

MAINTENANCE
AND SERVICING

*I*f *you had the brain of a mole, throughout your lifetime you could* only use it as the neuronal circuits installed in it allow it to be used: for the life of a mole. And just by living the life of a mole, you would already be doing everything necessary for the maintenance and servicing of that brain.

But your brain is not hard wired the way a mole's brain is. You can apply it to a great number of highly varied tasks. If you want to, through years of practice under the tutelage of a Persian Sufi master, you can get your brain to the point where you can unflinchingly walk over hot coals or let arrows be stuck through your most sensitive body parts. With an Indian yogi, you can train your brain in such a way that by your conscious intention you can influence your respiration, your heartbeat, and a series of other bodily functions that are normally controlled autonomously by centers deep in the brain. You can bring these functions so completely under your will that if you show these abilities to your doctor, he will begin to doubt everything he has read in his Western textbooks about the autonomous regulation of bodily functions. With the Inuit, you could learn how to differentiate numerous different kinds of snow; and with the native peoples of the Amazon, you could learn to recognize and name over a hundred shades of the color green. If you wanted to, you could also become a juggler and get the movement-coordination abilities of your brain to the point where with two hands you can keep seven balls in the air at once. And if you cannot think of anything else, you also have the possibility of keeping your brain very fully occupied watching colorful images on television, solving crossword puzzles, playing computer games, or learning the phone book by heart. These occupations also use and stabilize particular neuronal connections.

Thus a human being, in contrast to a mole, is not only able to decide freely what he is going to use his brain for, but also what he wants to make out of it. When he makes a particular decision and resolutely sets about implementing it, he no longer really needs to concern himself with the maintenance and servicing of his brain. He just has to refrain from letting himself be diverted from the path he has chosen. Just by being used from that time on only for the purpose he has decided upon, the inner organization of his brain will become better and better adjusted to the performance now required of it. Where there's a will, there's a way, and if the will is strong enough and the same way continues to be traveled, it gradually turns into a major road and maybe even into a freeway—this happens in the brain too. And because later on it becomes increasingly difficult to get off of these well-established highways, the decision a person makes concerning how and for what purposes he uses his brain is one that he should only make with a great deal of circumspection and care.

For example, it would not be very clever to make such a decision based on temporary possibilities and demands, because it has to be sustainable over the long term and take into account at least all the future developments that are foreseeable. Such a decision should still remain appropriate when a person has grown older and his needs have begun to be different. And it should not be something that hinders him later on when his feeling, thinking, and behavior have to adapt to the new demands of a world that will inevitably change.

It would also be shortsighted for a person to decide how and for what purposes he is going to use his brain purely on the basis of the conditions, possibilities, and needs that are current where he is living, that is, in a particular family, in a particular small-town or city community, in any particular cultural setting in any particular time. No one can rule out the possibility that later on he might have to move away or that times will change and with them the circumstances that he finds in a particular place. If that occurs, the same thing is quite liable to happen to him as would happen to a mole if for some reason it found itself among the sweet-smelling flowers on a sunny meadow; or as happened to Konrad Lorenz's goose, who because it grew up with him, thought for the rest of its life that the old man's world was the only proper goose world, and that anyone who looked like Lorenz was a goose like itself.

Neither geese nor moles will greatly envy us our freedom of

choice. They can usually rely quite well on what has been programmed into their brains. We humans, on the other hand, have a brain that to a certain extent programs itself on the basis of the way it is used. So we have to decide how and for what purposes we are going to use it. Now if a person comes to the conclusion that he is not going to make any such decision at all, then the final state of neuronal connectivity in his brain will automatically be determined by his genetic predispositions and by the conditions in which he grows up and lives. Thus he will remain a prisoner of his passively adopted, built-in tendencies and of the circumstances he happens to encounter. On the other hand, if a person decides to use his brain in a very particular way to attain a particular end, then he runs the risk that the inner organization of his brain will keep adapting itself better and better to this kind of one-sided use. In that way, he will increasingly become a prisoner of his own decision.

The point is, we can only remain free by deciding as early on as possible and as prudently as possible how and for what purposes we are going to use our brains. But for a number of reasons that is more complicated and much more demanding than we think.

To take an example, no person can freely decide how and for what purposes he is going to use his brain if he is suffering from hunger, cold, or some other physical hardship or if he is subject to psychological torment of some kind. The same is also true for people who are continually in a state of fear over whether what they possess—their wealth, their power, and their influence—is going to be taken away from them by people who do not have these things. Moreover, people whose thinking and behavior are ruled purely by their feelings are no more capable of making a free decision about how to use their brains than people who let themselves be driven entirely by their intellect and continually suppress their feelings. And finally, nobody can make a free decision about how and for what purposes he is going to use his brain if he does not have the faintest idea what is going on his brain and what the different ways are in which it could be structured and used. This goes for people who up to now have not had the chance to acquire this knowledge, as well as for people who have been flooded with such a surplus of information for such a long period of time that at some point they have not only completely lost any overview they might have had but also the ability to distinguish what is important from what is not important and what is false from what is true.

Thus a prudent and careful choice about the use of the brain, one that takes into account everything important that has already happened and everything important that might still happen in the future, cannot be made either just from the gut or just from the head, and definitely cannot be made as long as either one of those two is either too full or too empty.

Not only every individual human being but also all human beings from a particular cultural milieu pass in the course of their development through an initial phase in which the feeling in their gut is stronger than intellect in their head. In the long term a human brain can only come up with one solution for satisfying those strong demands coming from the gut and that is to use and develop the intellect more. When people begin to use their heads more to satisfy their gut-level demands, sooner or later they come to the realization that they can do this better collectively than they can alone. But then, all the people involved have to be in agreement about which feelings are the top-priority ones they have to deal with and about what strategies for satisfying these feelings they see as the most promising and likely to succeed. The more complete their agreement is, the greater their joint efforts will be, and then it is only a matter of time before the envisaged goal will in fact be attained. At that point, those who wanted enough to eat will be satisfied. Those who felt threatened by others either will have built protective walls around themselves or subdued whatever they had seen as threatening. And those who wanted to live a comfortable life will be sitting around in their cushy living rooms. The belly will be full, but the common goal that guided these people for such a long time in using their brains will be gone. Their common quest for shared solutions will have reached an end. From this point on, each of them will once again begin pursuing his own individual path.

Much of what was built up in the previous stage now begins to fall apart and be forgotten. Important experiences that were accumulated over generations by these people on their way to the attainment of their common goal, experiences that forcefully made their mark on the brains of earlier generations and then on those of their offspring, are now no longer available and begin to lose all the importance and standing they once had. There is a general loss of orientation, a sense of groundlessness. Since the old thinking no longer has much application in the new circumstances, sooner or later a feeling of discontent appears. If the deterioration of society reaches the point where once again the most basic needs are no longer being met, where people are once again being threatened by enemies, or where the basic comforts of life have ceased to be there, then the whole game starts over again from square one. A feeling shared by all determines a common goal, and when it has been achieved, everything falls apart again.

But not quite entirely. For each time, something is left over from this apparently senseless cycle. There is a little more knowledge, a few special skills, and a few new experiences. Perhaps some of these skills and some of this knowledge came from formerly alien cultures with which the people in question came in contact in the course of their earlier common quest. And if all this knowledge and the skills arising from it and all these experiences and the realizations deriving from them are not destroyed or rendered useless in the course of subsequent spells of groundlessness and disorientation, then the treasure hoards of experience accumulated over many generations by a great many different cultures can begin to expand and blend together. This process occurs in stages. There are progressive levels of perception, of knowledge, and of consciousness. At each one of these stages, new possibilities for a more comprehensive and more complex use of the brain open up. Progressively better conditions are created for the fulfillment of the potential that exists for developing a really human brain.

Under particularly favorable circumstances, the people of a particular culture might manage to make an especially big move upward on this ladder. But these great moments in history do not go unbalanced. Conditions also continually appear that lead people to fall from complex but unstable levels of brain use they have attained in the past back to simpler but more stable ones. Even in the case of highly favorable circumstances, it is always initially only a few pioneers who achieve the leap up to more complex levels, which they then make workable for the many who follow. And even in the case of highly unfavorable circumstances, there are always a few individuals who are not willing to follow the others in the step down to a simpler level of perception, knowledge, and consciousness.

What makes these farsighted, sensible, and courageous people stand out from the rest is not their appearance, their power, or their influence, but the way in which they use their brains—in the most total and comprehensive way possible. For what they are seeking is not something specific and definite but simply the most that is possible. And since this goal can never be reached, they create a path, a way, to this unreachable goal.

5.1   On the Ladder of Perception

Our textbooks tell us that human beings have six senses. They can use them to see, smell, taste, touch, hear, and to tell when they have lost their balance. On the basis of these perceptions, we are able to deal with the external world and develop a mental representation of how it is constituted, how it can be changed, and whether or not danger is threatening us from out there. This representation is stored in the brain in the form of specific neuronal connective patterns. The representation that is composed of all these sense impressions is not, of course, a true picture of the actual nature of the external world; it is merely the picture that we, with all our limitations, are capable of making of this world. We can only see light of certain wavelengths, hear sounds of certain frequencies, and we cannot smell, taste, and touch everything that is out there. We can only register those things that have turned out to be important for the survival and reproduction of our species over the course of evolution. In spite of these limitations, what we are able to learn of the world outside through our senses is usually enough to survive in that world and even occasionally take pleasure in it.

We seldom become conscious of the fact that we perceive a whole additional range of signals from our inner world that we use to regulate our inner order. Changes in blood sugar levels; in concentrations of oxygen and carbon dioxide; in body temperature; muscle tone; circulation; in the activity of our inner organs and the signal substances, hormones, and mediators they produce—without our noticing it, our brain also perceives all this and a lot more of what goes on in our body. In this way it continually puts together a picture of what is going on in us. And whenever something in this picture of our inner world starts going awry and threatens to go completely off the charts, the brain instigates a counterreaction to try to restore our inner order to its original state.

This also we usually do not notice. Only sometimes, when the disturbance to the inner order gets particularly serious, does the counterreaction set in motion by our brain become somewhat stronger and more conspicuous. Then we have the feeling that something is the matter. We gasp for air (because oxygen is lacking); we have gas (because we have eaten something indigestible); we are hungry and feel queasy or even dizzy (because our blood sugar level has sunk too low); we get gooseflesh and the shivers, or we start to sweat and turn down the heat (because our body temperature has gone either down or up); we get thirsty (because the salt concentration in our blood is no longer right); we get in bed (because we are exhausted); we feel the need for sex (because our testosterone level has gone up); or we lose this inclination (because we are afraid and stress hormones are being secreted that drive our testosterone level down); or we feel an irresistible appetite for sweets or fried food (because the metabolic activity resulting from eating them triggers changes in our brain that have a calming effect).

So our brain is capable of perceiving not only what is going on in the external world that might be threatening, but also what is going on inside us that might threaten our normal inner-world order. And when this inner world falls into disarray, the brain instigates responses and reactions to quell the disturbances that have arisen. All this is nothing special. All brains do it, including the brains of animals. Here we are talking about the lowest, most primitive level of our perceptual capacity.

What animals are not nearly so good at as we are is the art of evaluating perceptions, of ascribing greater or lesser significance to them. We are able to view specific changes in our outer—but also our inner—worlds as noteworthy. Since we frequently and intensively activate the neuronal circuits that participate in registering, processing, and storing these kinds of changes, these circuits are especially well worked out and are easier to activate than others. As a result we can perceive and relate to certain specific phenomena better and faster than others. We are more or less sensitized to certain perceptions. We have our senses sharpened in a very specific way.

But we are also masters when it comes to dumbing down our senses. At first consciously—and later unconsciously when the neuronal connections necessary for this have become sufficiently well burned in—we suppress certain perceptions. Sooner or later this usually leads to drastic consequences. Our ability to pay a high level of attention to some things and not others has been the source of many a great discovery in the course of human history, but also of many a false alarm. Certain individuals developed this knack to the point where they could see things that all other people were blind to, and they could feel and sense changes coming of which the rest of the people had not the faintest inkling. And along with these perception specialists and prophets, there have also always been people who can supposedly "hear the grass growing," and try to predict the future from the positions of the stars and from the burnt bones of goats.

What distinguishes the real prophets and seers from the phony ones is the fact that in the course of their development they were able to sharpen all their senses simultaneously, not only those used to perceive changes in the external world, but also those used to perceive what was going on inside them. They developed the ability to use all these senses at the same time and in balance with each other. In so doing, they attained the highest level of perception of which the human brain is capable. The only people who can reach this supreme level are those who during their lives continually find a balance between emotion and intellect, dependence and autonomy, and openness and self-differentiation. To sharpen his senses in this manner, a person has to learn both how to grasp and how to let go. He must develop the ability to devote himself thoroughly and fully to a particular perception, to take it in completely, and to sense what it causes to take place within him. And he must let the inner image that arises as this happens fuse with all the other images that are already there inside him, so that they make one whole integrated picture, which is then really more like a feeling. When this happens, he must not allow himself to become so aroused by this feeling that he becomes identified with it and loses himself in it; rather he must be able to detach himself from it yet nevertheless preserve it within him from then on. Only if he relates to it in this way, will he subsequently be able to take in further new perceptions, both from his outer and inner worlds via other senses with the same intensity; feel what happens inside him as he does so; and then fit these new "feeling images" together with all the other previously stored ones into an ever more comprehensive picture of his outer and inner reality.

All of us were able to do this, at least on a rudimentary level, when we were children. Many of us, however, have lost the ability. Those who have lost it only rarely feel anything when they perceive a change in their outer or inner world, and when they do, it involves only a few, faint images that certain feelings still arouse in them. However, it is possible to restore this ability that has been lost but nevertheless remains characteristic of a human brain. Instead of continuing just to scan the world perfunctorily or to look at it through a very narrow optic, it is possible to couple particular images, smells, or sounds with feelings, to really allow what is happening out there to enter us, and to actively connect these new impressions with all the other images that are in us from the past. It is possible to retrain oneself in this. It can be practiced. But for this, leisure time is required as well as stable inner balance, an undisturbed environment, and a resolute will. Whoever lacks the last of these and cannot find the first will inevitably continue to have his perceptual capacity determined by those circumstances that always compel him to use particular senses in a very particular way. In that case his perceptual capacity will adjust without any conscious participation of his own, automatically, so to speak, to this habitual manner of using his senses. Such a step down on the ladder of perception happens all by itself. You can only go up if you want to. And to want to take such a step up, you need a reason.

5.2 On the Ladder of Feelings

When the brain perceives a change in the external world or in the world of our body that leads to an upset in the balance of the hitherto harmonious flow of information processing taking place in the brain, a feeling arises. This feeling tells us that something out there in the world around us or something within us is not right. Most often we experience a feeling of this sort when we perceive something that does not fit in with what we were expecting, when demands are made on us that we cannot fulfill, or when someone hurts, deceives, or cheats us. We have lots of names for this feeling: insecurity, despair, impotence, helplessness. But whether we like it or not, whatever we call it, it remains what it is: fear.

And then, when somehow or other we have succeeded in restoring the inner order in our brain, and thus at the same time in our body, we also perceive that as a feeling that we give different names to: hope, satisfaction, confidence, sometimes even pleasure. But again here, these are only different names for that other basic feeling, the one that always makes its appearance when we have managed to conquer our fear: joy.

And there is a third basic feeling, which makes its appearance at moments when we cannot assess precisely whether what we have perceived should be seen as a threat to our inner order or as an opportunity to recover and stabilize our inner order: surprise.

There is no reason for us to think that other kinds of beings who possess a brain do not know and feel these three feelings. Just as we can, all other animals living in social groups can communicate these feelings to their fellows, perhaps by producing and emanating a particular odor, by carrying out particular bodily movements or adopting certain postures, by producing certain sounds, by calling out or screaming at times of danger, or by grunting, blabbering, or purring with contentment. The ones that have faces that can convey expressions have the additional ability to express their feelings through a characteristic facial gesture.

This language of feelings is generally understood by all other members of a given species and especially well by members of the same family or clan. It is the most important instrument of intra-species communication, and therefore it is especially highly developed in those species whose survival is most particularly dependent on skills like recognizing threats very quickly so they can be warded off through group action or making known newly discovered share-able resources so they can be secured through group action. This language of feelings also became highly developed among species with a critical need to strengthen and stabilize the emotional bonds between members of families, clans, and other groupings.

The ability to communicate feelings, not so much through odors but mainly through gestures, mimes, and uttered sounds, must have played a key role during the phase of transition to the human level. The gift of being able to express certain feelings is thus still laid by us in the cradle at the time of birth in the form of certain genetically programmed neuronal circuits in our brains. In addition the ability to recognize particularly important feelings in other people, such as fear, joy (pleasure), disgust, sadness, and pain, is also genetically configured in our brains.

These gifts are not equally well developed in all newborns. And what becomes of them—whether they are further developed and elaborated or are suppressed and deteriorate—depends on the conditions a child grows up in. Children can be brought to perceive their feelings in a more or less subtle fashion, and to express them more or less forcefully. Children who are trapped in situations of insecure bonding learn with astonishing speed not to show their feelings, to hide them, or even to express feelings that in reality they do not feel at all, but that they know are expected from them in certain situations. Many people develop in this way into masters at the game of playing with their own feelings and those of others. They are often quite cunning at this and observe other people very keenly. But they lack the ability to put themselves in the place of other people and empathize with their feelings. They are perfect masters at the keyboard of displaying simple basic feelings, but they are incapable of developing them into more refined and subtle feelings.

Such people are pretty well stuck on the lowest rungs of the ladder of human feeling capacity. Their feeling, and thus their thinking and behavior as well, is primarily determined by self-centered considerations. Consequently they themselves are the ones who reduce and narrow their ability to feel. In order to break beyond these limits, people of this type must be given the opportunity to enter into close emotional relationships with other people again. This is the only way they can gain the experience that these kinds of relationships can provide security and that within that realm of security it is possible to let one's own world of feeling merge with another person's. They have to learn again that not only is it not dangerous but also it is enriching to put oneself in another person's place and be able to feel what is going on in that person.

Empathy requires a tremendously refined level of perceiving and processing other people's nonverbally expressed feelings. The capacity for empathy can only be developed by people who are willing, and possess the necessary sensitivity, to place themselves within another person's world of feelings. It is this capacity that sets the human brain apart from all other nervous systems. The more thoroughly this capacity is developed and the more intensively it is used to enter through one's own feelings into the inner world of not only one other person but many others (and even into the inner worlds of other kinds of creatures), the higher a person can climb on the ladder of human feeling.

5.3   On the Ladder of Knowledge

Basically speaking, a nervous system has no other job to do besides defending against, or compensating for, changes in the outer world that might lead to disturbances of the inner order of an organism. Progressive optimization of the structure and function of the nervous system so it can manage this job more and more efficiently has inevitably led, over inconceivably long periods of time, to the formation of brains that allow their possessors to perceive threats to their inner order very early on, to evaluate the effects on themselves of changes in the outer world increasingly well, and to react with increasing specificity to these threats.

Thus from what began as strictly genetically programmed structures arose structures that were programmable through the individual experiences of the possessor of the nervous system in the early stages of life, and then later on, to structures that were programmable through the experiences of the possessor of the nervous system throughout his or her life. In the course of this development, the level of complexity and the extent of networking of the neuronal connections in the brain have continued to increase, but the basic way the brain functions has changed very little. Just as an individual nerve cell conducts an impulse only when it is strongly enough stimulated by impulses coming to it from other nerve cells, the brain also only instigates a compensative regulatory response when a perceived change in the outer or inner world is major enough to stimulate neuronal networks lying deep within the brain. This activation of limbic centers is something we experience as a disturbance to our emotional balance. What we pay particular attention to, what particularly excites us, how we evaluate a perceived change, and how we finally react to it, all depend on the experiences we have had in our life so far with disturbances of this sort. Many of these experiences are so general in nature that in the course of human phylogenetic history they have already, through selection, taken the form of genetic programs that produce very specific patterns of neuronal connectivity in our brains. Other experiences become anchored in our brains only through our having them in our own lives. Most of our own formative experiences—and usually the most important of them—occur in our early childhood, without our ever thinking about them or even being able to grasp them in words. For this reason, they are unconscious and often remain so throughout our lives.

Brains that are capable of initial learning or even lifelong learning are advantageous (and therefore came into existence at some point), because with such a brain an individual can take in experiences—initially or even throughout life—that are crucial for survival and reproduction. The ability to use such a brain to *know* consciously what is happening in and around its possessor was a relatively late development. Up to the present time, this capacity for conscious knowledge has arisen only to a rudimentary degree in a few of our nearest animal relatives, and in ourselves has reached only a certain level.

This special capability makes it possible for many apes and most humans to derive a general "if-then" kind of knowledge from all the unconscious experiences they have accumulated. This represents the lowest and most primitive level of knowledge. The basic recognition that certain effects are traceable to certain causes is something every child is capable of. Once it has experienced such a recognition, it continues to seek out new causal relationships in its perceived world. Where in its world it finds such relationships depends to a great extent on the people who help the child in its exploration. These people determine how far it can go in its climb up the ladder of knowledge. Many primary caregivers succumb to the temptation to direct the child's attention particularly, or even exclusively, toward causal connections in the external world: "If you flip this switch, then the light will go on." "If the fuse is out or the power company isn't producing current or the circuit has been broken, then it doesn't go on." And finally, "Electrical current is produced when. . . ." In this way, nowadays, every child learns to trace certain observable phenomena in the external world back to specific causes. We owe all our knowledge of the cause-effect relationships in the world around us to the ability developed in this way.

Success makes us blind, and excessive canalization of our thinking into cause-and-effect relationships has its price. People who stay stuck on this level of cognition sooner or later come to regard the whole world as knowable, and also everything that they have come to know as a simple cause-effect relationship as doable. This holds for violent criminals as well as for unscrupulous businessmen, politicians, and scientists.

At some point, however, most people discover that the majority of the phenomena of the external world come about through a number of causes working together. And so normally every person arrives some time or other at the (often painful) knowledge that a certain cause that he or she has set in motion in order to achieve a particular effect has set off a whole chain reaction that produced effects that he or she did not foresee.

Then from this knowledge arises the further knowledge that the perceived phenomena of the outer world are the result of interactions that are complex, hard to comprehend, and often unpredictable. On this level, the recognition of complex, mutually conditioning relationships appears. Every person who has reached this stage inevitably comes to see himself as limited in the freedom of action he previously thought he enjoyed. From then on, in order to get better at assessing the unintentional consequences of his actions, such a person must work with ever greater care and circumspection. There are people who consider themselves "big shots," who will not tolerate such limitations on their freedom of action and who therefore choose to remain standing on the first rung of the ladder of knowledge. It is not unusual for such people to cause considerable harm through their one-sided, goal-oriented actions.

All others, once they have noted the unintended consequences of actions of theirs, have to ask themselves if they want to continue behaving as they have. Such people are now on the third and highest rung of the ladder of knowledge, the rung of self-knowledge.

This rung is most easily reached by those who have had occasion early on in their lives to take notice of the effects of their outward-oriented actions on themselves, on their bodies, and on their brains. Most of these people understand fairly soon that everything one does leaves a trace—including in oneself. This is both a painful and a wholesome cognition, one that only a human brain is capable of.

5.4   On the Ladder of Consciousness

In recent years scientists in the field of brain research have been providing more and more compelling evidence that all our behavior, our highest rational functions as well as our emotional reactions, are based on certain neuronal processing activities that go on in our brains. Highly complex activities such as perception, memory, planning, decision making, and even intuitive feeling and evaluation depend on an equally complex foundation that is at once tremendously intricately networked, yet material. This holds true also for the all-important attainment that is generally held to distinguish human beings from animals: consciousness.

By consciousness we mean the ability to be aware of our own feelings and perceptions, our "being-in-the-world." Here the primary processing activities on which the brain's functions depend themselves become the object of cognitive processes, and the results of this meta-analysis are represented once again on a higher level. In order to develop consciousness, the brain must be able, so to speak, to observe itself. By building up meta-levels on which internal processes are reflected and analyzed, a brain can arrive at the point of being conscious of its own perceptions and intentions. It can grasp the state of what it has become and its role and place in the world. This ability has been developed to different degrees by different people. What level of consciousness a particular person can reach is inextricably bound up with how high he has been able to climb in the course of his life on the ladders of perception, feeling, and knowledge.

Typically, both on the level of human history as a whole and on the level of the personal history of any individual, the ladder of consciousness begins with the appearance, out of a dreamlike state of concrete identity with the life of the body, of a small kernel of inner experience that grows and becomes progressively clearer and more autonomous. With the emergence of this experience the primal stage of mythical consciousness is left behind. Through a process of step-by-step detachment from an original close bond with nature (the natural environment, early caregivers) arises both the possibility and the necessity of thinking about oneself. The emergence of this individual consciousness is at the same time an awakening out of a paradisiacal feeling of unity with the world. At this stage, a person begins to see himself as an autonomous, free, independently deciding and evaluating ego.

This process of transition is a difficult one that has yet to reach its end point in many cultures even today. There are always certain individuals who are the first to make the leap from the primal collective mythical stage of consciousness to the level of ego-related (self-) consciousness. Cultural and intellectual-historical evidence tells us that this transformation of consciousness began in the so-called Western cultural world about six thousand years ago. The first clear expression of it comes in the Gilgamesh epic, the tale of the heroic deeds and the personal life of the king of Uruk, written more than three thousand years ago. It took until the beginning of the Enlightenment for enough people to reach this stage of consciousness of their own egos so that this could become the basis of the prevailing (average) consciousness in the cultural world of the West.

As this ego-oriented (self-) consciousness became more and more widespread among the population, the period of time during which children could remain on the level of mythical consciousness decreased. For many of today's children, a slow and gradual onset of the process of becoming conscious of their own ego and their role and place in the world is a thing of the past. A growing number of children now quickly develop a kind of pseudoautonomous self-centeredness that, in its many and varied manifestations, has come to represent a serious threat to the stability of Western society.

This errant development makes it clear how important it is for a person for a consciousness of his own to grow and mature gradually in and from himself. When a person has a certain view of himself and his place in the world forced or imposed on him by the circumstances in which he grows up, various attitudes and convictions arise, but no real consciousness of his own develops. It is true that with these attitudes and convictions he can live and deal with the world, but he will be unable to take full advantage of the potential of his human brain, that is, to become conscious of himself, of the state of what he has become, and of his "being-in-the-world."

Worse still, a person who passes through the phase of mythical consciousness in only a very abbreviated and one-dimensional manner will subsequently barely be able to develop from within himself an autonomous, self-reflecting ego consciousness at all. Without such a consciousness of his own, he will remain pretty much imprisoned by (and dependent upon) ideas that he has taken over from other people in an unconscious and unreflective manner. In relation to our metaphor of the ladder of consciousness, he will more or less fall off all the rungs. He will be programmed by others and thus will be subject to their manipulation.

It will be much the same for people who grow up in a cultural and psychological environment that prevents them from discovering their own ego. In many languages—Chinese, for example—there is no word at all for what we, entirely as a matter of course, call "I." In such cases, the individual can only describe and understand himself by representing his relationship to others. What can all too easily come about in such circumstances is an unreflective, collective consciousness that hinders the individual as well as his ego-consciousness from developing their potential.

Despite the strong forces at work to canalize the thinking of members of a community in one direction or another, a certain number of individuals have always succeeded in freeing themselves from the concrete circumstantial pressure of prevailing opinions and attitudes and managed to develop a general, all-encompassing conception of humanity and its place in the world. This is what is called transcendence, and the level of consciousness attained by it is transcendent (or transpersonal, or cosmic) consciousness. At the present time it is hard to imagine that at some point all human beings will reach this highest level of consciousness. But the fact that it has been attained again and again by individuals already makes it clear in principle that a human brain—and only a human brain—is capable of it.

5.5   *Practical Advice*

Having established the only direction a human brain can really take on the ladders of perception, feeling, knowledge, and consciousness, two practical questions remain open.

The first is: Why should a person take the trouble to embark on this difficult path? Why should he sharpen his senses and try to perceive changes in his outer and inner worlds as sensitively and precisely as possible? Why should he develop the capacity to put himself in the place of other people and to empathize with their feelings? Why should he try to know himself and ultimately even to become conscious of what is taking place in himself, conscious of who he is and how he has become what he is?

The answer to this first question is simple. If you take a difficult path, you begin to use your brain in a significantly more complex, multifaceted, and intensive fashion than somebody who complacently remains until his last gasp where he has either ended up accidentally or been dumped by the push and pull of circumstances. The type and intensity of brain use determines how many connections are built up among the billions of nerve cells in it, what patterns of neuronal connectivity become stabilized there, and in how complex a fashion these neuronal connective patterns interconnect with each other. Thus in making a decision about how and for what purposes you are going to use your brain, you are also making a decision about what kind of brain you are going to end up with. This may be an unpleasant and uncomfortable realization, but that is simply the way the brain works. We didn't come to have brains capable of lifelong learning just so that we could set ourselves up comfortably in life. We possess them rather, so that with their help, we can take steps on the path of development, not only at the beginning of our lives, but throughout them. Of course we are always free to choose to stay where we are at any given point, and from that point on to use only the neuronal circuits that have already been established in our brain. But the more frequently we use these circuits in the same old way, the better and more efficiently set up and worn in they become, and so that the choice to just stay as we are could very well end up being the last free decision we ever make in our lives. Once we have gone ahead and successfully programmed our brains for a very specific kind of use, then as long as nothing else intervenes, the rest runs by itself, to the very end. By then, the chance to put in place program-opening structures, the chance for the comprehensive use and complex configuration proper to a human brain, has passed us by.

If you do not want to stay stuck in your well-worn ruts of perception, feeling, and knowledge and thus lose your freedom, you have to choose the hard path and try—rung by rung on the ladders of perception, feeling, knowledge, and consciousness—to come closer to that which distinguishes a human brain from all others: the ability to keep calling itself into question, again and again.

This brings us to the second practical question. How does one achieve this ability and how does one hold onto it? Certainly not, as has been suggested in the media lately in such glowing terms (claiming to represent the latest findings of brain research), by occasionally going down the stairs with your eyes closed, smelling a flower, or surprising your colleagues with a new behavior pattern or a novel hairstyle. Just deciding from time to time to do something you ordinarily do not do does not bring about changes in the neuronal circuits in your brain. To really change the circuits, we must create conditions that will not only make it possible but actually urgently necessary to perceive more of what is going on around us, to feel these perceptions more thoroughly and deeply, to evaluate them in a more complex fashion, and above all, to think about them more carefully before we decide to do one thing and drop something else.

There are only two routes we can take to bring about these kinds of conditions, one comfortable and one uncomfortable. The comfortable route is the one we are already familiar with, the one regarding which we have already had the opportunity to accumulate a rich store of experience in the course of our development thus far. This is the path on which we simply just try to keep going, with all our mistakes and limitations. Unfortunately, as time goes along this path gets more and more tiresome, until we finally get completely hung up in the tangle of all the problems our limited approach creates. Only when it becomes impossible to keep going along on this path just as we have been doing, do we finally reach the insight that the way we have been using our brain has failed us. To call ourselves into question in this way is not only quite painful but also quite dangerous, especially if we have taken other people along with us on this path, and on top of that, if it has seemed to us for a long time that we were making really good progress on it. Success makes us blind, and communal success all too easily also blinds the people who are really the most open and see the best. Those people are the children who grow up in the community. With the help of their enormously flexible and learning-capable brains, they are in an excellent position to take over all the capabilities and skills, ideas and convictions of the people they grow up among. Of course, they most readily adopt whatever of all this seems the most critical to them for dealing with their lives. The more successfully the parental generation has progressed along a particular path by using a particular strategy, the more likely it is that their children will not only follow them on this path, but that later as adults they will lay out this path more efficiently and tread it with even greater resolve.

And since the more exclusively you concentrate on a specific goal, the better progress you make toward achieving it, these offspring will tend, even more than their parents did, to push away, not to perceive, or to suppress anything they take to be useless or cumbersome in achieving this goal as quickly and directly as possible.

Whatever the goal might be—assertion of personal interests; attainment of power and influence, of fame and recognition; the dominance of one's own tribe, people, or nation; the spread of a particular faith; the achievement of a political ideal; or the realization of a crazy idea—the results of these efforts is always the same, the path just varies in length. The more thoughtlessly a particular goal is pursued, the sooner one gets caught up in the tangle of problems that result from one's own shortsightedness and inattention. And if these consequences do not catch up with the fathers, they will catch up with their sons or grandsons. Sooner or later people will have to face the mess and ask themselves what has been the matter with the way they have been using their brains. In any case, they have become richer by one more experience. And through this, whether they wanted to or not, they have come one step closer to the second path, which begins where the first, the initially seemingly easy and comfortable one, so painfully ended— with the ability to call oneself and the way one has been using one's brain into question once more.

No one sets out on the other, more demanding path voluntarily. A person must feel compelled. And this path can only be entered upon if a person continually retests his behavioral patterns and attitudes toward himself and everything around him. The best approach is for the person to ask if what he considers truly important really is so important.

Once behavioral patterns and attitudes have taken us over, we are no more conscious of them than we are of the power they exercise over us in forcing us to use our brains in a particular way. Inattention, for example, is a behavioral pattern that does not require much in the way of "brains." If a person manages to be more attentive and careful, then, she will automatically put more "brains" into whatever she perceives, into whatever she associates with her perceptions (what she activates in the brain in connection with them), and into whatever she includes in her decision-making process than somebody who just keeps relating with himself and the things around him in a superficial and heedless manner. Thus attention—care—is a highly essential factor in the service and maintenance of a human brain.

What attention and care can achieve in terms of the fundamental expansion of brain use on the level of perception and psychological processing can be achieved on the level of the neuronal connections that are responsible for our decisions and behavior through an attitude that we call gentleness. Through a lack of gentleness, that is to say, thoughtlessness and inattention, a given goal can be achieved in a hurry. However, complex neuronal circuits are not needed for this approach. Such an approach neither uses nor firmly establishes any.

If one begins to think what basic attitudes one must adopt in order to use one's brain in a more comprehensive, more complex, and more highly networked fashion, a whole series of concepts come to mind, many of which have almost begun to disappear from our current vocabulary: sensibleness, uprightness, humility, prudence, truthfulness, reliability, courtesy. All these are basic attitudes that were regarded as worth aspiring to in times when brain re-searchers did not yet exist—(to say nothing of complicated picture-producing technologies like computer-aided positron emission tomography, with whose help nowadays we can compare the brains of an attentive and an inattentive person and demonstrate the differences that result from the different ways they have been used).

A person can no more develop these attitudes all by herself than she can learn to speak a particular language or to read or write a book all alone. For these things, she needs other people who can read and write who can demonstrate these skills. And what is still more important, she has to have a close emotional relationship with these people. They have to be important to her—just as they are, with everything they can do and know and also everything they cannot do and do not know. She has to like them not because they are especially good-looking or rich people, but because they are as they are. Children can be open to other people in this way and love them without reserve, just for themselves. For this reason children most easily take on the attitudes and speech of the people they love. And sometimes adults too can relate to each other just as unreservedly and selflessly as children do. Love creates a feeling of connectedness and solidarity that transcends the person loved. It is a feeling that keeps spreading outward until in the end it includes everything that brought us—and all the people we love—into the world and holds us here. A person who loves in this expansive way, without reserve, feels connected with all things, and everything that is around him is important to him. He loves life and takes pleasure in the multiplicity and colorfulness of this world. He enjoys the beauty of a meadow glistening with morning dew as well as a poem that describes it or a song that sings it. He feels a deep awe before everything that lives and that life brings forth, and he is sorely moved when any part of this is destroyed. He is curious about what there is to discover in this world, but it would never occur to him to take it apart out of pure greed for knowledge. He is grateful for what nature has given him. He can accept it, but he does not wish to possess it. All he needs are other people with whom he can share his perceptions, his feelings, his experiences, and his knowledge. A person who wishes to use his brain in the most comprehensive manner must learn to love.

6

WHAT
TO DO
IN CASE OF
MALFUNCTION

*I*n the case of a mole's brain, we recognize a malfunction when the brain no longer does what it has to do for the mole to lead a proper mole's life. When the mole no longer knows where it should burrow, when it loses his way in its tunnels and even begins to confuse repugnant-tasting roots with edible earthworms—and at such a point it is not helped by anyone—the mole is lost.

People are not moles. When something goes wrong with their brains, usually somebody comes along who tries to help them. People are most easily drawn to offer this kind of assistance in cases where the malfunction is quite pronounced, especially if it is life threatening. But finding help turns out to be a lot harder if the malfunction in question is only a so-called partial-performance deficit. These are malfunctions that create a situation where one can still do a great deal but can no longer do certain things, so they are disturbances in which the brain is still functioning almost normally. Even in cases like these, a person can usually find somebody to help. However, finding help is hardest of all for people whose brain is still functioning quite normally but who are leading a life that is anything but human. This has been the case in the past for people who spent their lives as Roman slave dealers in Egypt, Spanish plunderers of the native peoples of Peru, or scalp collectors among the Indians of North America. Things were no better for people in the more recent past who put their brains to work as Nazi thugs in Auschwitz, poison-gas manufacturers in the town of Leverkusen, or mercenaries in the jungles of Vietnam. And of course this also applies even today to people who are using their brains as arms dealers, child molesters, environmental polluters, speculators, liars, fencers of stolen goods, and cheaters. Anybody who constantly uses his brain to achieve his own interests at the expense of other people—and is thus feeling, thinking, and behaving in anything but a human fashion—finds getting somebody to help him lead a more human life extremely difficult. For him, things are not really that much different than they are for a mole with a malfunctioning mole's brain.

In spite of this, such people remain alive and often even live longer than people who have human brains and who for that reason persist in using them in a human fashion. This is hard to understand. Either somebody is helping these people to survive, or what we call *human being* is not a biological appellation for a definite species that has reached a definable and stable stage in its development. Both of these are true. The process of becoming human has still not reached its conclusion, and we have obviously far from fully exhausted the possibilities for the development and application of our brains. We are still en route, already half human and still half animal; we remain undetermined, we are still searching. For this reason, we are also ready to accept as human and take on as companions anybody who looks like we do and who possesses a brain that in principle ought to be just as capable of learning as ours.

We can only really take on other people as companions if we know what path we want to travel together. Once we have decided to follow the path that leads toward greater humanity, we can try to approach this goal through common effort. That is the point at which it really makes sense to come to grips with those malfunctions that prevent an otherwise normal brain from being used as a *human* brain. At that point it is important to be able to recognize this kind of malfunction as early as possible, with the first signs and symptoms. Becoming human is an extraordinarily complicated and therefore very fragile process, in the course of which we continually run the risk that a distortion of the process resulting from malfunction will be declared normal. At that point, of course, the question of where we should be going ceases to apply.

6.1   User Errors

Not all who think they see something have their eyes open; and not all who look about know what is happening around them and with themselves. Some only begin to see when there is nothing left to see anymore. Only after they have already brought their house and home down around them do they start to become

sensible people. Seeing how things are too late brings no remedy,
only sorrow.

—BALTASAR GRACIÁN

We do not owe this advice concerning a very frequent error that can be made in the use of the brain to modern brain research. It dates from the seventeenth century and appears in the *Oraculo manual*, a text composed by the Jesuit priest Baltasar Gracián (1601–1658) as a kind of handbook or mirror of self-knowledge. Along with shortsightedness and blindness, Gracián describes in this book a host of other user's errors, which in general lead to a failure to benefit fully from the many-sided potential that exists for the use of the human brain. These errors include complacency, arrogance, indolence, superficiality, bias and narrow-mindedness, thoughtlessness, and (yet again) inattention.

If you look around today, you can plainly see that Gracián's advice has not been put to much use. This also holds true for the insights of other wise guides who have held up before our eyes the limitations of human perception and thought, making use of more or less witty, cryptic, or sometimes, cynical images to do so. We look at these books, are amused by the shortsightedness and folly people manifest in the use of their brains, but then we do no more than blithely gloat over the simplemindedness and stupidity of others. The moment it comes down to seeing our own limitations in these images, the entertainment quickly ends. The better polished the mirror of self-knowledge is and the more clearly and undeniably it reflects the errors we make in the use of our own brains, the sooner we lose our desire to keep looking at ourselves in it.

That which is obvious is not always easy to understand. This is especially true regarding the really important things in life. A thing only really becomes important to a person when he himself is personally touched by it and it makes him feel a sense of deep personal concern. Deep personal concern always arises when a person has to acknowledge having made a mistake. It is a profoundly unpleasant feeling because it calls into question and challenges the way we have been thinking, feeling, and behaving. It not only forces us to look at ourselves but also forces us to change. And the less willing we are to change, the less able we are to comprehend the mistakes we have been making in the use of our brains, as obvious as they might be. For this reason, most people, operating with ego-centered, shortsighted, one-sided, superficial, and thoughtless strategies, have to experience failure or breakdown on some level before they can get a look at themselves and understand the mistakes they have been making. "Only after they have already brought house and home down around them do they start to become sensible people," said Gracián. And he also noticed that: "Many recast an unsuccessful enterprise as a moral duty; having taken the wrong path, they see it as strength of character to keep going on it."

But there has been a great and significant change since Gracián's time. In his days it was mostly individuals who brought house and home—and sometimes whole principalities and kingdoms—down around their heads as consequences of their limitations. In our times, however, a multitude of individuals has turned into an anonymous mass entity, and the many individuals comprising this mass entity, now endowed with collective blindness, are in the process of bringing the house and home of all of us down around our heads—on a global scale. They foul the air; alter the climate; pollute rivers, lakes, and seas; destroy the natural habitat; and squander the earth's resources. They stand by and watch as more and more people lose the basic underpinnings of existence, as the rich variety of natural life forms and human cultures dwindles, as rainforests are cut down, oceans are fished out, and fertile lands are turned into desert. They see all this plainly. Newspapers and television parade it before their eyes on a daily basis. But somehow they do not really feel a sense of deep personal concern about it. And as long as all these people manage to ward off and suppress the feeling of deep personal concern, they can and will go on behaving as they have, using their brains in the same old way.

Anybody can make a mistake. In fact we cannot avoid making mistakes again and again. Only by doing something wrong can we learn how to do it right. A person who made no errors in the use of his brain would not be able to change. He would be like one of those robots that are optimally programmed to perform specific tasks but is incapable of further development. But a person who keeps managing to suppress the feeling of deep personal concern regarding his mistakes, and who is able to stave off all doubts concerning the rightness of his thinking and behavior, at the same time robs himself of his chance to learn. He loses the ability to correct his mistakes, to change, to develop further. He becomes more and more like a lifeless and unfeeling robot. Thus he has lost precisely that which characterizes a brain as human—the ability to step out of well worn ruts, to undo already existing programming. Consequently, suppressing and resisting the feeling of deep personal concern is the only real error a user can make in using his brain.

A person can be too shortsighted in his thinking, too narrow in his perceptions, and make inattentive, superficial, complacent, narrow-minded, thoughtless, and otherwise limited use of his brain. However, as long as he is still capable, in the face of all these errors and inadequacies, of developing a feeling of deep personal concern and self-doubt, he is also still capable of changing. However, if he successfully manages to keep suppressing this feeling, he can and will continue to use his brain as before, until he has pulled house and home down around his head.

The insidious part of this user's error is the fact that deep personal concern and self-doubt are extremely unpleasant feelings. No one happily and willingly calls himself into question in this way. A person will all too readily grasp at every chance that comes up to stave off this kind of insecurity. This work of suppression goes best if he can submerge himself in an anonymous mass composed of many other people, whose desires, hopes, and fears he can share. On top of this, since time immemorial, the desires, hopes, and fears of these many other individual people have proved to be wonderful things to exploit in achieving one's own personal ends. There have always been individuals who have seen this possibility with particular clarity and taken advantage of it skillfully to achieve security, stability, power, influence, wealth, and prestige for themselves. People who are successful with this strategy have the least reason to doubt the rightness of their thinking and behavior or to feel deep personal concern—even in those cases where the way in which they have achieved their success really ought to trigger precisely that.

Deep personal concern can be really well suppressed if a person is able to set a very high value on himself, his goals, and his ideas, and to regard himself as more important, more on the mark, and on the whole just generally superior to other people who have different goals, attitudes, and convictions. The more like-minded people he can find, the easier this is for him. When these like-minded people add up to a very large number, it is only a matter of time before the other "inferior" people are declared the common foe of the nation and are persecuted and killed. This is done with conviction, without any individual sense of deep personal concern, and without any doubt as to the correctness of one's own behavior. For deep personal concern can only be felt by a person when he destroys something or sees something destroyed that is important to himself. And something can only be important to a person if he feels a sense of close connection to it. Otherwise it leaves him cold.

It takes no great art to use and influence the human brain in such a way that it eventually loses the capacity to arouse or let in a feeling like deep personal concern. In the second half of the last century, we learned to master this art, to the extent that it is one, and transmit it to our children as never before in the whole history of humanity. The basic approach is quite simple: All that has to be done is to ensure that nothing else is really important to a person besides living the most comfortable life possible. For this to work, the person must be prevented from developing close bonds with other people, with his home, with nature, and with everything that surrounds him. He must not put down any fixed roots and he must not notice the fact that with his clipped wings he can no longer fly. He must be kept in a state of continual excitement with trivial matters, be flooded with useless information, and confronted with so many expert opinions that he can no longer distinguish important from unimportant or true from false. To prevent him from reflecting seriously, it is advisable to keep him rushing about frantically until he loses the ability to sit still for more than five minutes at a time, to say anything meaningful, or even to think about what he is going to do next. You can also overstimulate his brain with lurid and exciting images, loud and shrill noises, and continual sensational input, until his ability to perceive has been completely blunted. And if you keep him in a continual state of agitation with fresh reports of catastrophes and images of brutal violence and inhuman crimes, at some point his ability to feel will also die.

The earlier on you can provide a person with all these possibilities and cause him to use his brain in the corresponding fashion—that is, the more malleable his brain remains—the surer you can be of achieving the desired results. And if, in spite of this, the inevitable occurs and some individuals realize that what is going on around them does implicate them personally, and when consequently a feeling like deep personal concern does arise in their brains and they begin to doubt the rightness of their attitudes and convictions, then all you have to do is persuade them that everything is under control, it is all being taken care of, things are still workable, and it can all still be fixed. There is nothing people would rather believe than that. They gratefully grasp at every straw that will enable them to escape the sense of personal implication and compunction arising from their dark depths and get back into the same old habitual flow.

Relieved, they buy a ticket for a quick trip to Honolulu or go a shopping junket to London, Paris, or New York. They rent the latest horror movie or watch the constantly replenished scare news on TV. They surf the Net for hours without knowing or acknowledging to themselves what they are looking for there, or they log onto a chat group and exchange trivialities with people they have never met and never wish to meet. They read in the newspaper that there are people who keep having cosmetic surgery until their faces have turned into grotesque masks or that others have rings put around their stomachs because they cannot control their appetites and have gotten so fat they can hardly walk. And it amazes them that there are doctors who perform such operations and journalists who make their living writing about them. They get hold of all the pills and drugs that they have heard touted for the relief of their problems or the increase of their pleasure; or they simply reach for the liquor bottle when their frustration gets to be too much for them. They are in favor of less traffic on the roads, but they buy all kinds of products that have to be shipped to them from a distance, shrimp that are caught in the North Sea and shelled in Morocco, carrots grown in Germany and washed in Sicily. They spend their time as observers, watchers. They click through the channels on TV or lose themselves leafing through magazines. Then they complain that they have so little time.

They are continually on the lookout for opinions that confirm that they are absolutely just fine the way they are. And they are grateful when they find an expert who, through his objective, scientific findings, definitively proves the correctness of their convictions, opinions, and attitudes. And they maybe even buy a user's manual for their brain and read it with the expectation of finding tips in it for how to protect what is supposedly their most important organ from user's errors without significantly changing the way they have been using it up to now.

6.2 *Error Messages and Damage Control*

In a technical apparatus, errors in usage result in the apparatus not functioning the way it is supposed to. Often they lead to a breakdown of some kind, and if your luck is running bad, through a small error you can turn your whole expensive machine into a pile of junk. The same goes for a brain. You can cut off its blood supply (through strangulation), you can stop its oxygen supply (through suffocation), or so disrupt its functioning (through poisoning) that it can never be restored. Usually a brain reacts to such interventions with such alarming error messages that the perpetrator stops before it is too late. Only someone who has lost his faith in the worth of the functioning of his or any other human brain can ignore these warning signals.

In fairly complicated technical apparatuses, such as computers, protective mechanisms are built in to protect the device against serious errors in usage. Then there are the moderate user's errors that, though they do not destroy the computer, result in its not doing what it should do, or in the operator not being able to make use of all its potential functionality. A person who cannot operate his computer properly inevitably ends up reducing it to a commodious typewriter or a somewhat complicated Game Boy and begins to regard it as no more than that.

The same is essentially true with regard to the brain. The only difference is that, in the case of the brain, the mechanism does not remain as it was but gradually takes on the form of what it is used for. Just like a computer, the brain does not sound any alarm to make us aware of the fact that we are in the process of reducing it, out of ignorance of its inherent potential or mere complacency, to the thing we are continually using it as. Basically, the brain does not care any more about this than a computer does. As long as it keeps being able—even as the stunted version of what it could be—to perceive all threatening changes in the outer or inner world in a timely fashion and to compensate for them, it doesn't make a peep. It only sends out an alarm when it is no longer capable of doing this, because the processing activities going on within it have gone haywire. Usually only when our brain responds to a threat to its inner order with a massive fear and stress reaction do we finally become aware that something has gone wrong. Many people, however, react even to this kind of emergency signal from their brains with a mere shrug of the shoulders and simply try to go on as before. Until they get physically or psychologically sick. That is the very last emergency brake the brain can apply. If a person is unable to regard even that as a chance to make changes in the use he has been making of his brain, then he has come to the end of his (and eventually all medical) possibilities of damage control.

In order to get out of such narrow, deep ruts, people need the help and support of other people, especially those who think, feel, and behave differently than they do themselves. The more complex the fashion in which a person's brain is networked with the brains of other people, the less danger there is that his individual user's errors will go unnoticed. And thus the likelihood is greatly increased that a well-networked person will be able to fully benefit from the multifaceted potential inherent in every human brain.

A computer behaves in a similar fashion. Anyone who has

connected his own computer to a complex computer network has had an opportunity to appreciate the many new possibilities this opens up for the use of the machine. But even in setting up and configuring such a network, there are pitfalls. Such networks are all too vulnerable to individuals or groups who might want to begin to use them as handy instruments for the achievement of their own specific ends and the spread of their own ideas. If they turn the whole network into a mere tool, instead of fulfilling its potential to offer a wide range of possibilities, it ends up being used for one primary purpose—manipulating the thinking, feeling, and behavior of everyone connected with it.

We can try to guard against this danger by setting up our computer network—or any other means of communication through which people influence each other—in such a way that each person can use it the way he wants. Then everyone has the chance to put forward and promote whatever he likes. What might come out of this is unpredictable. But what cannot come out of it is the very thing that a human brain needs to unfold and develop its many-sided potential. What a person needs to develop his brain is not the greatest possible number of relationships with the greatest possible number of other people that will enable him to exchange the greatest possible number of ideas or products with them. Rather what he needs is perhaps just a few—but intense and meaningful—encounters with other individuals of the kind that will make it possible for him to merge the various experiences he has had the opportunity to acquire in his life thus far with an increasingly larger and more inclusive bank of experience.

If more and more people simply pass each other by with no more than a superficial exchange and all their brains begin to adapt to this sort of use, then even when, as a result, the whole society's shared bank of experience becomes extremely fragile and gradually begins to collapse, no alarm will go off. Nothing will happen—at least nothing will happen as long as the whole system continues to function to some degree.

What can happen to a whole society is not that different from what happens to a single individual who manages, throughout his life, to deal with the whole range of his problems with one and the same strategy of action. Just like the individual, the society progressively loses its flexibility and creativity. It too becomes increasingly unreceptive toward whatever disturbs it in the execution of its hitherto successful strategy. It too finally breaks as a consequence of brittleness if it cannot get out of the ruts it has been traveling in and find new, more appropriate solutions for the problems it has created for itself. The individual has to reorganize the neuronal circuitry in his brain. The society has to reorganize the inner structures that determine the thinking, feeling, and behavior of its members.

These inner structures are actually not terribly hard to see and describe. On the bottommost layer, deeply anchored, we find the attitudes and convictions encountered and adopted during childhood, with all the more or less clear traces in thinking, feeling, and behavior left behind by the parental house and school. There, too, are the conceptions concerning what life is all about assimilated from peers, adults, and the media. On top of this foundation are piled all the further experiences that the developing person acquires during his education and his working life in dealing with the world he has inherited. Built into this is everything that has proven useful over time, that is, that has aided the individual in finding security and inner stability.

The most appropriate strategy, the most effective means, for achieving inner stability and security—perceived as by far the most important these days, and therefore most loudly proclaimed by many—is achieving psychological and material independence through the acquisition of power and wealth, or—if that does not work—at least by being able to display symbols of status indicative of power and wealth.

There is another way, one that has lost a great deal of its popularity in recent years but that is also well suited for coping with people's fears and for creating a sense of security: the acquisition of knowledge and skill. However, this strategy must inevitably dwindle in value in a society where the knowledge of the individual is being drowned in a tremendous flood of information and individual capabilities and skills are being replaced by computer-guided machines, which are causing an increasing number of people who are well equipped with experience and skills to sit around unemployed.

The third path a person can take to a sense of well-being and security in life is the path of social bonding, the anchoring of the individual within the community. This path can be found only by those who have had the experience in their lives of being themselves only a part of a larger whole. These people realize that they can attain security only by contributing to the cohesion and sense of togetherness of the community. Unfortunately, this is a path that nowadays is chosen by only a few people, and, what is more unfortunate, by very few people of influence.

A person must have as many different experiences as possible in his life with other people and in this way acquire such broad and comprehensive knowledge and such a wide variety of abilities and skills that he can no longer be made dependent, either materially or psychologically, on other people. This is the only way he can arrive at a place where he is able to choose freely how and for what purposes he will use his brain. But even such a person can change himself and use his brain differently only if he has recognized some decision he has made in the past as an error, and if he has felt deep personal concern as a result. For an entire society to be able to change, many individuals must feel this sense of deep personal concern. Thus to enter upon a different path, a given individual must know what he needs to pay more attention to in the future than he has in the past. For an entire society to enter upon a different path, many people must agree where they want to go together.

"We are the transitional stage between monkeys and human beings." This insight was imparted, many years ago now, by Konrad Lorenz as his gift to us for our journey. We still have the possibility of deciding where we really want to go—and of setting the appropriate example for our children.

6.3   Complaints and Liability

In case after reading this user's manual, you come to the conclusion that there have been shortcomings in your use of your brain thus far, you can regard the resulting feeling of uncertainty that begins to spread through your brain as a sure sign not only that you are alive, but that you do indeed possess a human brain.

In case such a feeling fails to set in, please, if you are still able, see your doctor or pharmacist, for as Gracián said, "He who does not understand anything* is not alive either."

---

*And, I would add, does not *feel* anything.—G. H.

Index